SCIENCE WORKS

CHEMISTRY
Mixtures and Solutions

Seymour Rosen

Upper Saddle River,
New Jersey

THE AUTHOR

Seymour Rosen received his B.A. and M.S. degrees from Brooklyn College. He taught science in the New York City School System for twenty-seven years. Mr. Rosen was also a contributing participant in a teacher-training program for the development of science curriculum for the New York City Board of Education.

Cover Photograph: Custom Medical Stock Photo/Chanes
Photo Researcher: Rhoda Sidney

Photo Credits:
p. 17, E: Helena Frost
p. 26: Lionel J-M Delivigne/Stock, Boston
p. 51, A: Helena Frost
p. 51, B: Helena Frost
p. 51, C: Rhoda Sidney
p. 51, D: Rhoda Sidney
p. 58, F: Helena Frost
p. 66: Don Kelly/Grant Heilman
p. 78, D: N.O.A.A.
p. 90, F: Gary Walts/The Image Works
p. 100, B: Gerard Fritz/Monkmeyer Press
p. 100, C: Rhoda Sidney
p. 100, D: Helena Frost
p. 100, E: Helena Frost
p. 102, F: Runk/Schoenberger/Grant Heilman
p. 106, C: Rhoda Sidney
p. 106, D: Arlene Collins/Monkmeyer Press
p. 106, E: Fredrik D. Bodin/Stock, Boston
p. 106, F: Grant Heilman
p. 123, A: Topham/The Image Works
p. 123, B: UPI/Bettmann Newsphotos
p. 125, E: Beaver Valley Nuclear Complex
p. 129, B: Reuters/Bettmann Newsphotos
p. 128, D: Reuters/Bettmann Newsphotos
p. 130, E: Beaver Valley Nuclear Complex
p. 131, F: Runk/Schoenberger/Grant Heilman
p. 138: Grant Heilman
p. 141, A: United Nations
p. 147, A: Hays/Monkmeyer Press
p. 147, B: Runk/Schoenberger/Grant Heilman
p. 147, C: United Nations
p. 147, D: United Nations
p. 162, B: Barry L. Runk
p. 162, C: W.H.O.
p. 162, D: H. Confer/The Image Works
p. 162, E: Hugh Rogers/Monkmeyer Press
p. 163, F: Philippe Plailly/SPL/Photo Researchers

ISBN: 0–8359–0329–0

Printed in the United States of America.
7 8 9 10 04 03 02 01 00

CONTENTS

MIXTURES

SOLUTIONS

ACIDS AND BASES

POLLUTION

Introduction to Mixtures and Solutions

Can you mix a pitcher of lemonade or make a milkshake? Lemonade and milkshakes are different kinds of mixtures. Just about everything in our lives—what we drink, even the air we breathe—is a mixture. Some things are "100% pure," but most things are mixtures. That is what this book is about.

You will learn the answers to some practical questions in this book. You will find out why antifreeze works in cars and why lemons are sour. You will learn why soda pop loses its "fizz" and why steam makes a teakettle whistle.

Finally, in this book, you will learn about the different kinds of pollution, how they affect the environment, and most importantly, how pollution affects you.

What is a mixture?

1

mixture: two or more substances that have been combined, but not chemically changed

LESSON 1 | What is a mixture?

Look around you. What do you see? Everything you see around you is matter. Matter is anything that has mass and volume. Volume is the amount of space matter takes up. Matter is not always visible. For example, the gasses in the air are matter.

As you probably know, matter is made up of tiny particles called atoms. There are more than 100 different kinds of atoms. Some matter has only one kind of atom. Matter that has only one kind of atom is called an element. Gold, copper, oxygen, and mercury are examples of elements.

Most matter is made up of two or more different kinds of atoms that are chemically combined. This kind of matter is called a compound. In a compound, atoms are linked together as molecules [MAHL-uh-kyoolz].

Pure water is a compound. Some other examples of compounds are salt, alcohol, sugar, and chalk.

Sometimes, elements and compounds are found "alone." They are not mixed with anything else. For example, the salt in a salt shaker is a compound. It is a salt "alone." No other substance is mixed in with it. If you added pepper, or anything else to the salt, it would no longer be salt "alone." It would become part of a mixture.

A **mixture** is made up of two or more elements or compounds that are not chemically combined. The parts of a mixture can be elements or compounds. And, these parts can be arranged in any way.

Many things you know are mixtures. Tossed salads, salad dressings, coffee, tea—even soft drinks—are mixtures.

As you know, matter can be in any state—a solid, a liquid, or a gas. Mixtures may be made up of solids only, liquids only, or gases only. In addition, mixtures may be made up of different combinations of solids, liquids, and gases.

The air we breathe and the water we drink are mixtures. Air is a mixture of gases. Drinking water has air and minerals mixed in with it.

ELEMENT, COMPOUND OR MIXTURE?

Figures A through E show some common substances. In the blank below each picture, write whether the arrow points to an **element, compound,** or **mixture.**

Figure A

1. ______________________________

Figure B

2. ______________________________

Figure C

3. ______________________________

Figure D

4. ______________________________

Figure E

5. ______________________________

WHAT KIND OF MIXTURE?

Six of the possible kinds of mixtures are:

a. a mixture of gases
b. a mixture of liquids
c. a mixture of solids
d. a mixture of gases in a liquid
e. a mixture of solids in a liquid
f. a mixture of solids and gases

In Figures F through K, each arrow points to a mixture described in the list above. Write the letter that shows what kind of mixture it is below each picture.

Figure F

1. ______________________

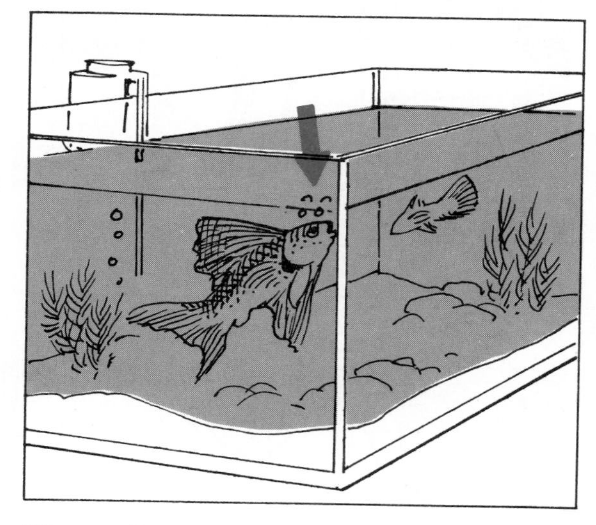

Figure G

2. ______________________

Figure H

3. ______________________

Figure I

4. ______________________

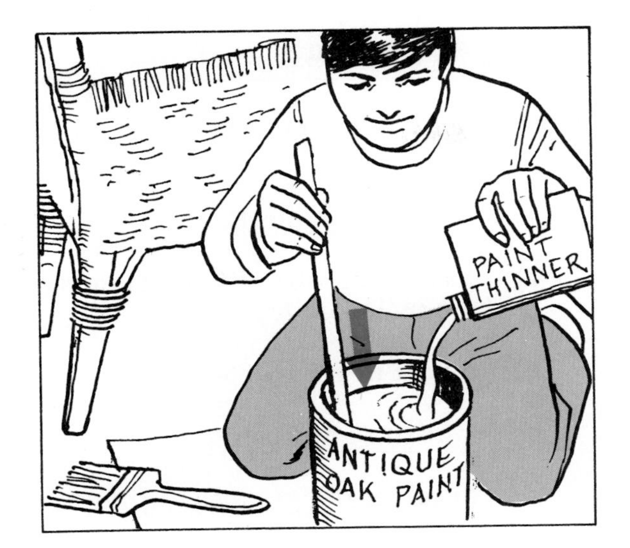

Figure J

5. ______________________________

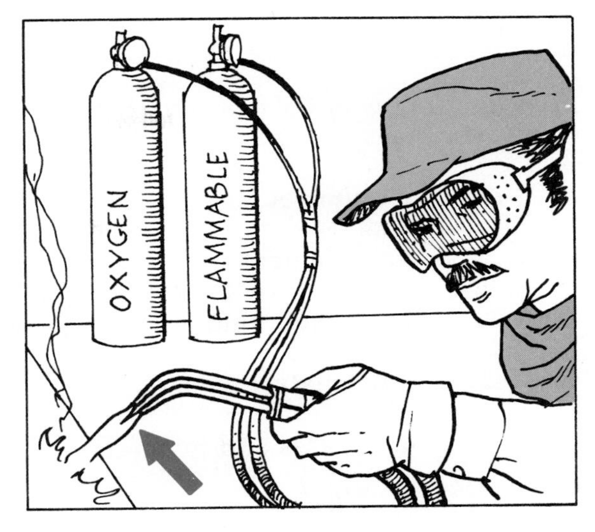

Figure K

6. ______________________________

FILL IN THE BLANK

Complete each statement using a term or terms from the list below. Write your answers in the spaces provided. Some words may be used more than once.

atoms	states of matter	element
liquid	solid	mixture
molecules	minerals	gas
gases	compound	air

1. The three states of matter are ______________, ______________, and ______________. (Use singular form.)
2. Every bit of matter is made up of tiny ______________.
3. Matter that has only one kind of atom is called an ______________.
4. Matter that has two or more different kinds of atoms linked together is called a ______________.
5. Atoms link together to form ______________.
6. Two or more different things close together is called a ______________.
7. A mixture may contain any combination of the ______________.
8. A mixture that is all around us is ______________.
9. Air is a mixture of ______________.
10. Water has several ______________ and ______________ mixed in it.

MATCHING

Match each term in Column A with its description in Column B. Write the correct letter in the space provided.

	Column A	Column B
______	**1.** element	**a)** has two or more different kinds of atoms linked together
______	**2.** compound	**b)** mixed in with water
______	**3.** mixture	**c)** a mixture of gases
______	**4.** gases and minerals	**d)** different things close together
______	**5.** air	**e)** has only one kind of atom

TRUE OR FALSE

In the space provided, write "true" if the sentence is true. Write "false" if the sentence is false.

______ **1.** All matter is made up of atoms.

______ **2.** A compound has only one kind of atom.

______ **3.** An element has only one kind of atom.

______ **4.** There are more compounds than elements.

______ **5.** Mixtures are made of substances that are alike.

______ **6.** Table salt is a mixture.

______ **7.** The air is a mixture.

______ **8.** Water is a mixture.

______ **9.** Oxygen is a compound.

______ **10.** We can see atoms.

What is a suspension?

2

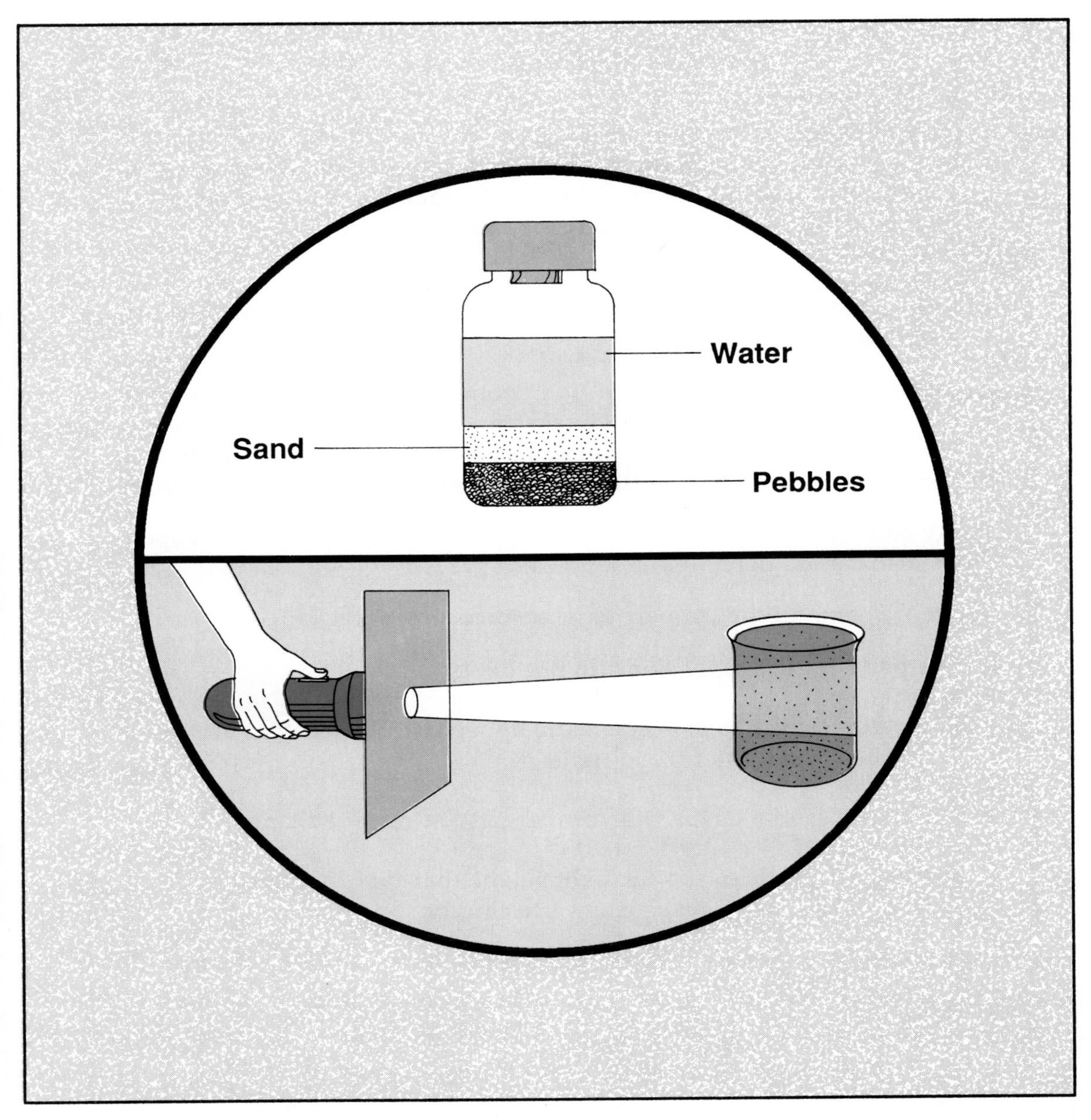

suspension [suh-SPEN-shun]: cloudy mixture of two or more substances that settle on standing

LESSON 2 | What is a suspension?

Before you pour orange juice, you shake it. Before you spoon out vegetable soup, you stir it. Orange juice and vegetable soup are mixtures. Mixtures such as orange juice and vegetable soup do not dissolve. The parts settle out.

Mixtures that do not dissolve and that settle are called **suspensions** [suh-SPEN-shunz].

You have many suspensions in your home. Salad dressings and fruit juices are suspensions. Look in your refrigerator. Some bottle labels may say "Shake well before using." These bottles contain suspensions. In fact, any mixture that you see settling or that needs mixing is a suspension.

Many common suspensions are made up of solids and liquids. Suspensions also can be made up of solids and gases.

What are some of the properties of suspensions?

- The particles in suspensions do not dissolve
- The particles of a suspension settle by weight. The heavy parts settle first. Then, the lighter parts settle.
- The particles of a suspension are large. You can see them easily.
- Suspended particles scatter light. Light that hits the particles is reflected. This is why suspensions are cloudy.

CLAY AND WATER

Figure A

Figure B

TRY THIS! Stir some powered clay into a jar of water. Let it stand. Notice what happens.

1. Powered clay in water ______________ a mixture.
   (is, is not)

2. The clay ______________ dissolve.
   (does, does not)

3. The clay pieces ______________ settle.
   (do, do not)

4. Clay in water makes a mixture called a ______________ .

5. The parts of a suspension ______________ dissolve.
   (do, do not)

6. The parts of a suspension ______________ settle out.
   (do, do not)

WHICH SETTLES FIRST?

Place some pebbles, sand, and powered clay into a jar (Figure C).

Add water nearly to the top.

Figure C

Figure D

Cover the jar tightly and shake.

Let it stand for five minutes. Observe what happens.

Figure E

1. Did the pieces settle in layers?

____

2. Complete Figure E. Show the layers. Label them.

3. Which pieces settled first? ______________

4. They are the ______________ (largest, smallest) pieces

5. They also are the ______________ (lightest, heaviest) pieces.

6. Which settled last? ______________

7. They are the ______________ (largest, smallest).

8. They also are the ______________ (lightest, heaviest).

9. This shows that when a suspension settles, the ______________ (heavy, light) pieces settle first; the ______________ (heavy, light) pieces settle last.

10. Usually, the heavy pieces are ______________ (small, large); the light pieces are ______________ (small, large).

TRUE OR FALSE

In the space provided, write "true" if the sentence is true. Write "false" if the sentence is false.

________ 1. Suspensions are mixtures.

________ 2. The parts of suspensions are dissolved.

________ 3. Suspensions settle out.

________ 4. Suspended pieces always settle at the same time.

________ 5. Suspended pieces settle by weight.

________ 6. In a suspension, heavy pieces settle last.

________ 7. Suspensions are evenly mixed.

________ 8. Suspensions are cloudy.

________ 9. Suspensions are transparent.

________ 10. Particles in suspension reflect light.

MATCHING

Match each term in Column A with its description in Column B. Write the correct letter in the space provided.

	Column A	Column B
________	1. mixture	a) reflect light
________	2. suspension	b) settle last
________	3. heavy pieces	c) cloudy mixture of two or more substances that settle on standing
________	4. light pieces	d) two or more substances that have been combined, but not chemically changed
________	5. particles in solution	e) settle first

HOW IS SEDIMENT DEPOSITED?

Figure F

Streams pick up and carry sediment such as sand, rocks, clay, and pebbles. Streams deposit this sediment where they empty.

1. Name some kinds of sediment streams carry.

   ______ ______ ______ ______

2. The ______ are deposited first because they are the ______ (heaviest, lightest).

3. The ______ is deposited last because it is the ______ (heaviest, lightest).

4. The ______ are deposited closest to the shore.

5. The ______ is deposited farthest from shore.

6. List the sediments in the order that they are deposited: ______,

   ______, ______, ______.

REACHING OUT

Sometimes, streams flood onto land areas. They leave sediment behind. What is this land good for? ______

______

______

How can the parts of a suspension be separated? 3

coagulation [koh-ag-yoo-LAY-shun]: use of chemicals to make the particles in a suspension clump together

filtration [fil-TRAY-shun]: separation of particles in a suspension by passing it through paper or other substances

LESSON 3 | How can the parts of a suspension be separated?

A cook drains off the cooking water from spaghetti. The wind makes dust fly in the air, but it settles. Spaghetti in water and dust in the air are both suspensions. In both examples, the parts of the suspension separated.

Many kinds of suspensions must be separated. Sometimes, nature separates the parts. Other times we must do it ourselves—or help nature along.

There are four ways to separate suspensions. They are **filtration** [fil-TRAY-shun], sedimentation [sed-uh-men-TAY-shun], spinning, and **coagulation** [koh-ag-yoo-LAY-shun].

FILTRATION Filtration is the same as straining. A filter has holes. Pieces smaller than the holes pass right through. Larger pieces are trapped by the filter.

Filters come in many sizes. Some have large holes. Some have small holes. The size of the filter you use depends upon the size of the particles you want to separate.

SEDIMENTATION Nature does this job itself. In sedimentation, the suspension just "sits." Gravity makes the pieces settle to the bottom of their containers.

SPINNING Spinning speeds settling. Spinning builds a strong outward force. The force pushes the pieces to the bottom of the container quickly.

COAGULATION Coagulation also speeds settling. Coagulation uses chemicals. The chemicals make small particles lump together. They become heavy and settle fast.

Coagulation occurs when you cut yourself. Chemicals in your blood cause the blood to coagulate and form a clot.

FILTRATION

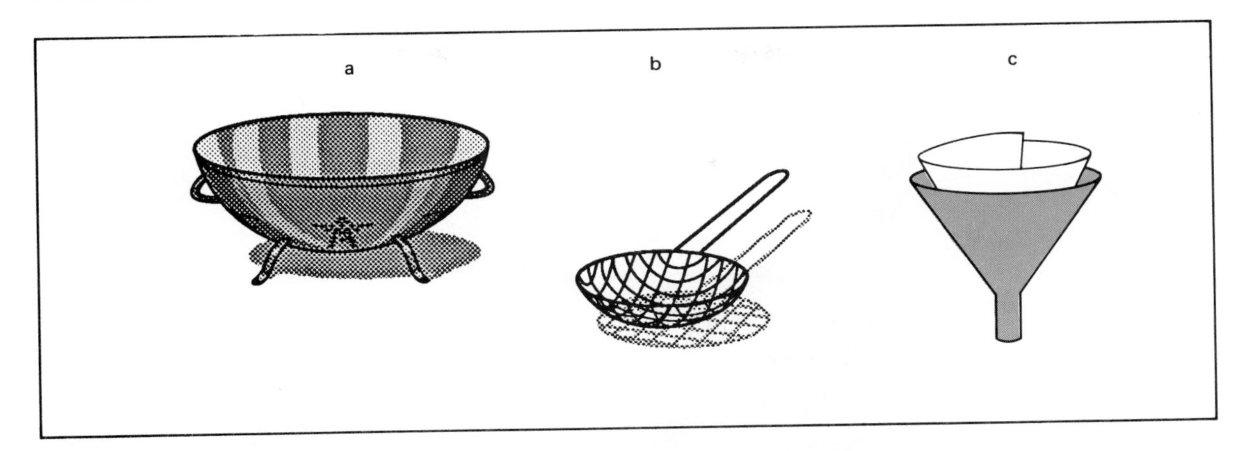

Figure A

1. Which filter separates out only the largest pieces? ________________

2. Which one separates out the smallest pieces? ________________

3. Which filter would you use to strain spaghetti? ________________

4. Does filtration use chemicals? ________________

SEDIMENTATION

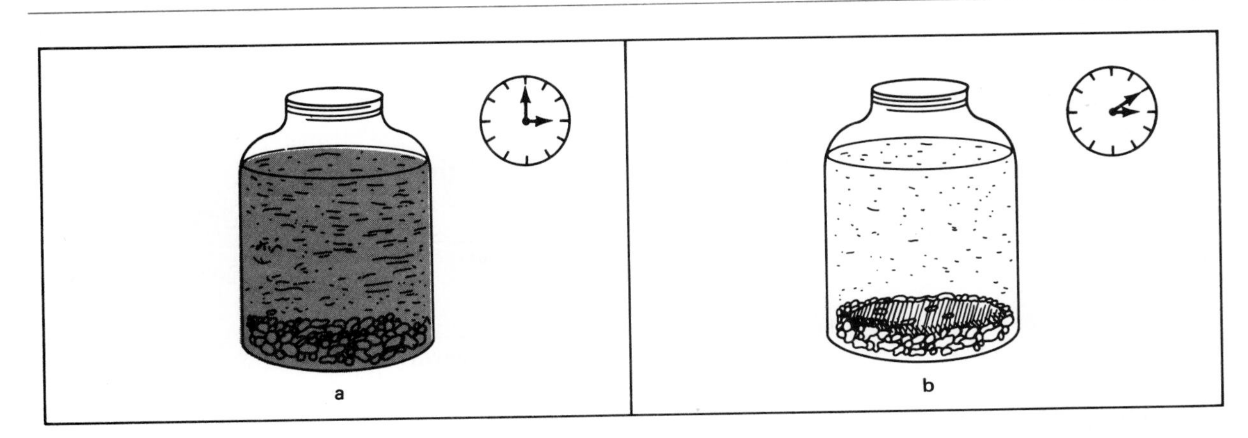

Figure B

1. In sedimentation:

   a) the heaviest pieces settle ________________ .
   first, last

   b) the lightest particles settle ________________ .
   first, last

2. Sedimentation is done by ________________ .
   gravity, spinning

3. Does sedimentation use chemicals? ________________

SPINNING

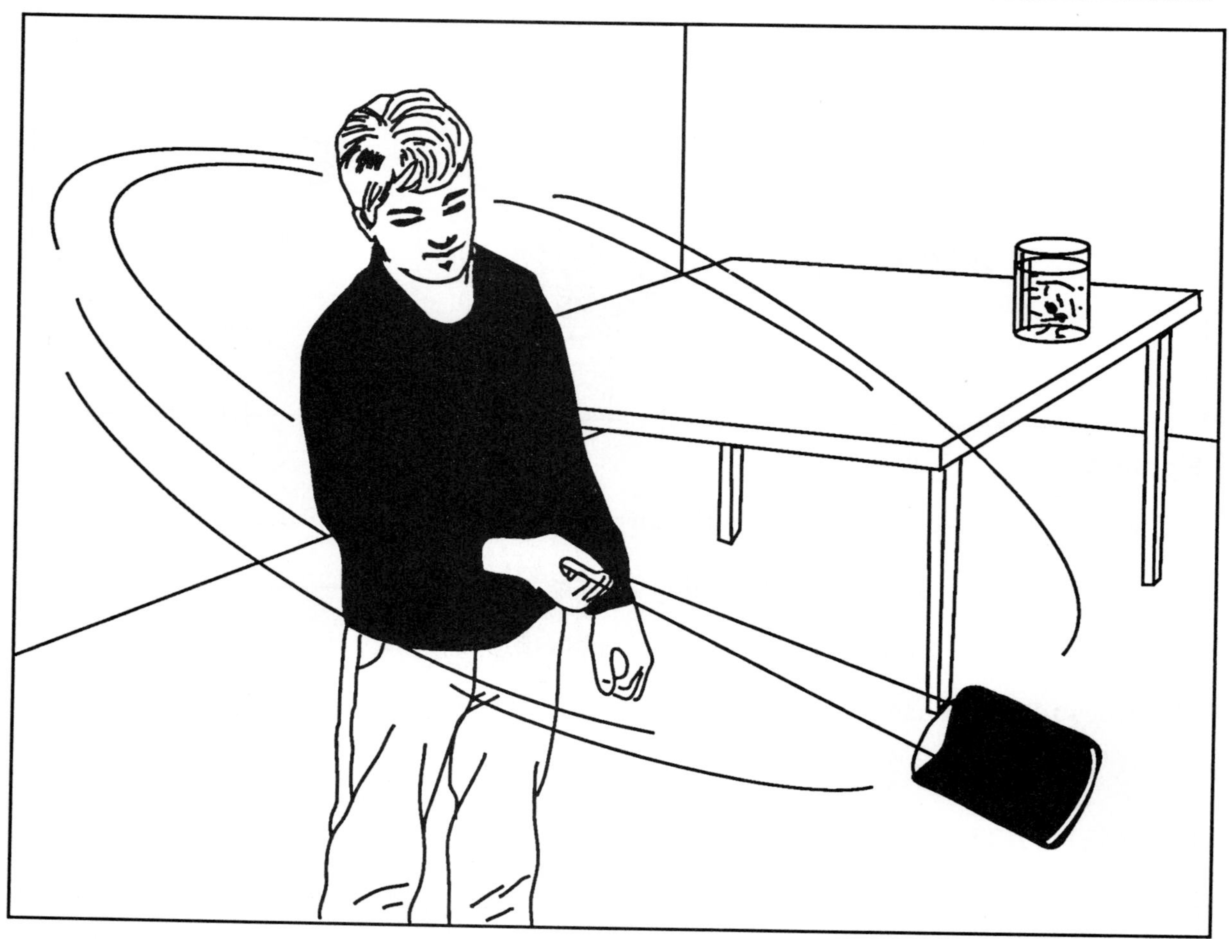

Figure C

There are suspensions of clay and water in the can and the beaker.

1. The clay pieces will settle first in the ______________ .
   beaker, can

2. The pieces in the beaker are settling by ______________ .
   spinning, gravity

3. The pieces in the can are settling by ______________ .
   spinning, gravity

4. Spinning causes an ______________ force.
   outward, inward

5. Spinning ______________ settling.
   speeds, slows

6. Does spinning use chemicals? ______________

COAGULATION

Figure D

1. The suspension in beaker b is settling by ______________ . (coagulation, sedimentation)

2. Coagulation makes small particles ______________ . (lump together, move apart)

3. Coagulated particles become ______________ . (heavier, lighter)

4. Heavy particles settle more ______________ than light ones. (slowly, quickly)

5. Coagulation uses ______________ and then gravity. (spinning, chemicals)

WHEN ARE SUSPENSIONS SEPARATED?

You have learned four ways of removing particles from suspensions. These can be very useful. Here is one example.

Most places get their drinking water from rivers, lakes, or reservoirs. The water has sediment suspended in it. It must be removed so the water can be fit to drink.

Several steps are used to remove the sediment. You will learn about these steps in Lesson 21.

Figure E

FILL IN THE BLANK

Complete each statement using a term or terms from the list below. Write your answers in the spaces provided. Some words may be used more than once.

spinning	first	coagulation
last	chemicals	lump together
gravity	sedimentation	filtration
outward	speed up	does not

1. The four ways of separating the parts of a suspension are ______________ , ______________ , ______________ , ______________ .
2. The method that separates a suspension by trapping particles is ______________ .
3. In sedimentation, ______________ makes the particles settle.
4. Heavy particles settle ______________ .
5. Lightweight particles settle ______________ .
6. Spinning and coagulation ______________ settling.
7. Spinning builds an ______________ force.
8. Spinning ______________ use chemicals.
9. Coagulation uses ______________ and then gravity to make particles settle.
10. Coagulation makes small pieces ______________ .

MATCHING

Match each term in Column A with its description in Column B. Write the correct letter in the space provided.

	Column A	Column B
________	1. gravity	a) causes strong outward force
________	2. filtration	b) do not dissolve
________	3. spinning	c) lumps particles together
________	4. suspension particles	d) pulls things down
________	5. coagulation	e) traps particles

TRUE OR FALSE

In the space provided, write "true" if the sentence is true. Write "false" if the sentence is false.

________ **1.** Suspended particles can be separated.

________ **2.** Filtration separates pieces by weight.

________ **3.** All filters are the same size.

________ **4.** Sedimentation uses gravity.

________ **5.** Spinning causes an inward force.

________ **6.** Spinning and coagulation speed sedimentation.

________ **7.** Filtration uses chemicals.

________ **8.** Spinning uses chemicals.

________ **9.** Sedimentation uses chemicals.

________ **10.** Coagulation uses chemicals.

WORD SCRAMBLE

Below are several scrambled words you have used in this Lesson. Unscramble the words and write your answers in the spaces provided.

1. NORFATILIT ________________

2. DOTRAWU ________________

3. NIPSNGIN ________________

4. TAGYVIR ________________

5. GALUNOTIAOC ________________

6. TATMENNIODESI ________________

7. CALCHEMSI ________________

COMPLETE THE CHART

Complete the chart by filling in the missing information. Identify whether each statement describes separation of a suspension by ***filtration, coagulation,*** *or* ***spinning,*** *by placing an X in the correct column.*

	Description	Filtration	Coagulation	Spinning
1.	Particles stick together.			
2.	Particles are caught on paper.			
3.	Motion causes particles to be pulled out of a suspension.			
4.	Chemicals are added to the suspension.			

REACHING OUT

After sediment from a suspension settles, how can you separate it from the liquid?

____

____

____

____

____

What are emulsions and colloids?

4

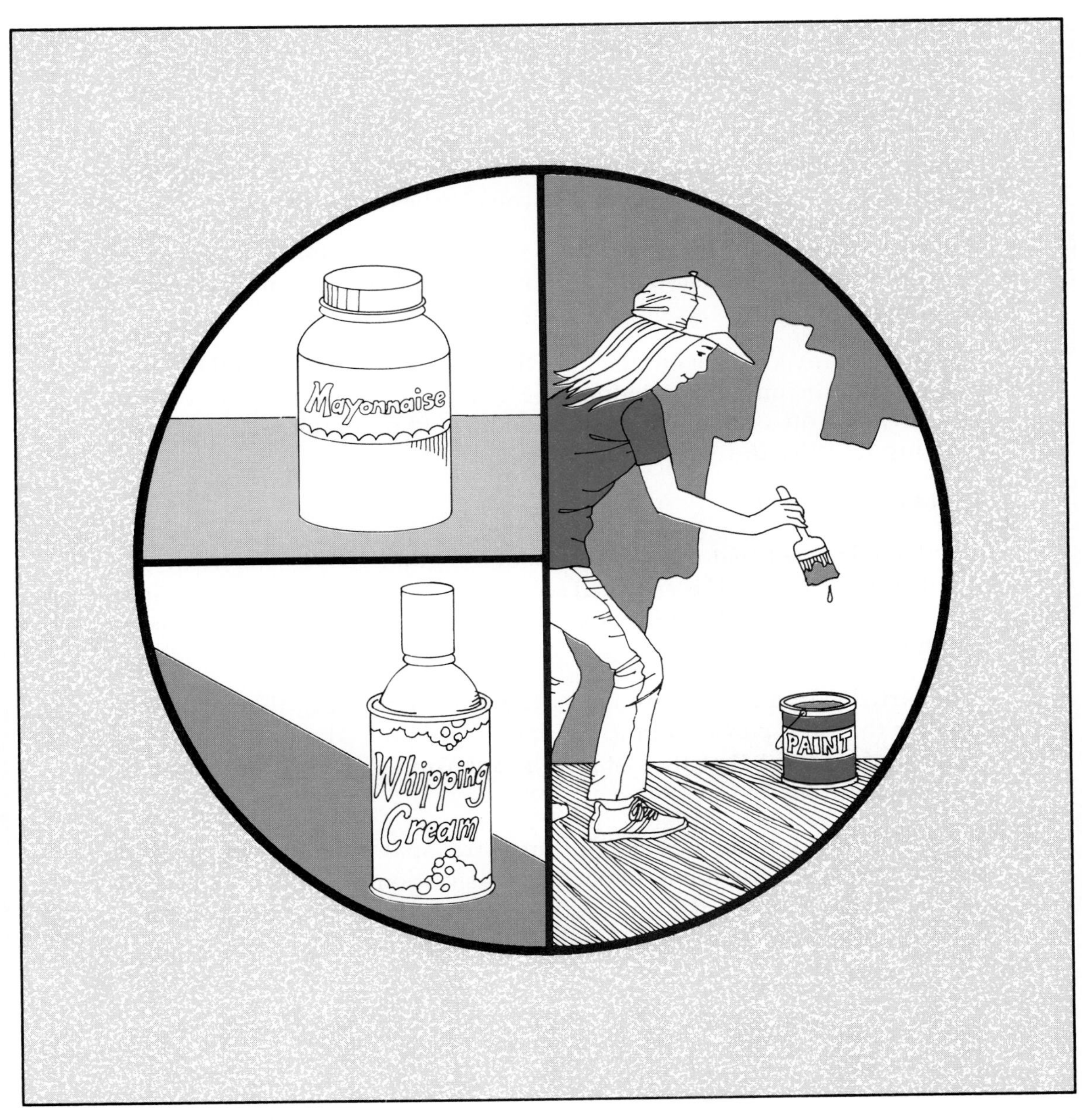

colloid [KAHL-oyd]: suspension in which the particles are permanently suspended
emulsion [i-MUL-shun]: suspension of two liquids

LESSON 4 What are emulsions and colloids?

EMULSIONS

You have seen mud settling out of water. Muddy water is a suspension of solids in a liquid. But suspensions can also be made up of only liquids.

A suspension made up of liquids is called an **emulsion** [i-MUL-shun]. The liquids in an emulsion do not mix evenly. They are immiscible (not mixable). The liquids separate into layers.

There are many kinds of emulsions. Oil and water is one example. A very common emulsion is oil and vinegar. Have you ever used oil and vinegar as a dressing on a salad? What did you have to do to the mixture before pouring it? Why did you do it?

COLLOIDS

You have learned that the solid parts of a regular suspension settle out. A **colloid** [KAHL-oyd] is a special kind of suspension. The solid particles in a colloid do not settle out. Fog, mayonnaise, and whipped cream are common colloids.

The particles in a colloid are larger than molecules. But they are much smaller than the particles in a regular suspension. The particles are so small and so light that they stay in suspension. They do not settle by themselves.

Most colloids look like liquid solutions—transparent and evenly mixed. You cannot see the suspended particles easily. Some colloid particles are so small that you need a microscope to see them. The particles in a colloid are always moving in random motion.

EXPERIMENTING WITH AN EMULSION

Figure A *Pour some oil and water into a jar.*

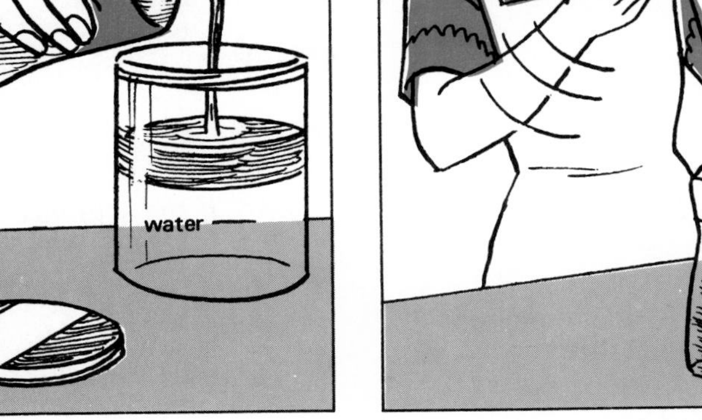

Figure B *Shake it very well.*

Figure C *Put the jar on the table. Observe what happens.*

Draw pictures of what you see . . .

after 30 seconds

after two minutes

after 10 minutes

1. The oil and water ________ mix evenly.
did, did not

2. The oil and water ________ separate.
did, did not

3. Oil and water ________ miscible.
are, are not

4. Oil and water make a ________ .
colloid, suspension

5. Oil and water is a special kind of suspension called ________ .
a colloid, an emulsion

TRUE OR FALSE

In the space provided, write "true" if the sentence is true. Write "false" if the sentence is false.

________ 1. An emulsion is a mixture.

________ 2. An emulsion is a suspension.

________ 3. An emulsion has solid particles.

________ 4. An emulsion has only liquids.

________ 5. The liquids of an emulsion are miscible.

________ 6. Colloid particles are the size of molecules.

________ 7. Colloid particles settle by themselves.

________ 8. Colloids look like solutions.

________ 9. Mayonnaise is a colloid.

________ 10. Colloid particles pass through regular filter paper.

MORE ABOUT COLLOIDS

TYPES OF COLLOIDS		
NAME	**PHASE**	**EXAMPLE**
Foam	gas in liquid	shaving cream, whipped cream
Emulsion	liquid in liquid	mayonnaise
Fog	liquid in gas	clouds, fog
Smoke	solid in gas	smoke in air
Gel	liquid in solid	butter, jelly

NOW TRY THIS

Using the information in this Lesson, identify whether each mixture shown is a colloid or an emulsion. Write **colloid** if it is a colloid; write **emulsion** if it is an emulsion.

Figure D

____

Figure E

____

Figure F

____

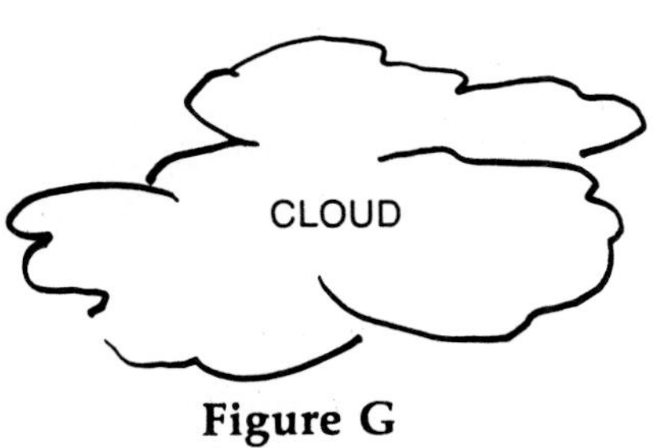

Figure G

____

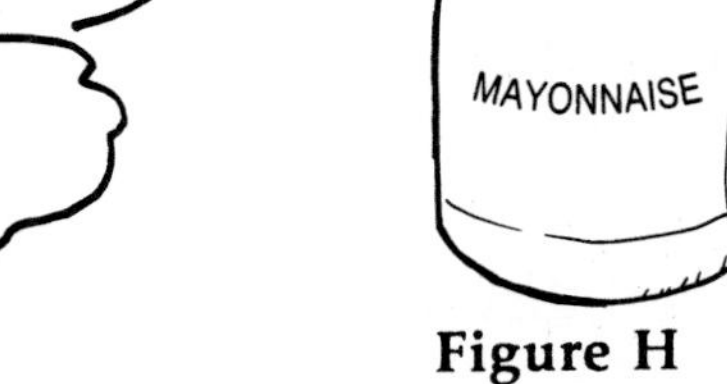

Figure H

____

SCIENCE *EXTRA*

Fluoridating the Water Supply

The fluoridation of public water supplies can help to reduce an expensive public health problem—tooth decay. Fluoridation is the addition of chemicals, called fluorides, to the water supply. In the 1930s, scientists discovered that people who grew up in the Southern part of the United States had fewer cavities than people living in other areas. They discovered that in these areas, the water was naturally fluoridated.

The scientists decided to see if the fluoridated water was the reason. They added fluoride to the water supplies in a few other cities. By the 1950s, the tests showed that the incidence of tooth decay had decreased in these cities. As a result, public health officials recommended fluoridation of the water supply for all communities. Today, about 60% of the water supply in the United States is fluoridated.

Fluoride is a trace mineral that occurs naturally in soil and water. Fluoride makes tooth enamel stronger. The stronger enamel makes the tooth more resistant to decay. Scientific studies have shown that a very small amount of fluoride results in a 50 to 60 percent drop in tooth decay. Fluoride is especially important before age six, when permanent teeth are formed.

The decision to fluoridate the water supply often is a controversial topic. The benefits of fluoridation, risks of fluoridation, and cost of fluoridation must be considered.

Large doses of fluorides can be harmful, especially to the bones and teeth. Recently, a study was published that stated that fluoride caused cancer in laboratory animals. However, this study was reviewed and it was found that the findings were not conclusive and more testing was needed.

Some scientists believe that fluoridation greatly affects people with kidney disease and those sensitive to toxic substances. However, these ill-effects of fluoridation have never been proven. Many health officials believe that the risk of harm from fluoridation is extremely small, and the benefits outweigh these risks.

Do you agree that water should be fluoridated? Is the water supply in your community fluoridated? If it is not, should it be? What do you think?

How can we keep emulsions from settling?

5

emulsifying [i-MUL-suh-fy-ing] **agent:** substances that keep an emulsion from separating
homogenization [huh-mahj-uh-ni-ZAY-shun]: formation of a permanent emulsion

LESSON 5 | How can we keep emulsions from settling?

The milk and mayonnaise you buy are emulsions. Yet they do not separate in layers. You do not have to stir or shake them.

This does not seem to agree with what you have learned about emulsions. But it really does. Certain things are done to milk and mayonnaise to keep them from settling. They are stabilized. Stabilized emulsions do not settle out.

How do you stabilize an emulsion? There are two ways:

- make the parts very small
- add special chemicals

MAKING THE PARTICLES SMALLER If you shake an emulsion hard, the parts break up. They become smaller. A blender makes the parts even smaller. Small parts take a long time to settle. The smaller the parts, the longer they take to settle.

ADDING SPECIAL CHEMICALS Most emulsions separate even if the particles are very tiny. They may separate slowly—but they still separate.

If we add special chemicals, they do not separate. Chemicals that stop emulsions from separating are called **emulsifying** [i-MUL-suh-fy-ing] **agents.** There are several kinds of emulsifying agents.

Egg yolk is one example of an emulsifying agent. It is used in mayonnaise. Mayonnaise is mostly oil and vinegar. Oil and vinegar are immiscible. They do not mix. A chemical in egg yolk keeps the oil and vinegar from separating.

Soaps and detergents are other emulsifying agents. They keep the grease that lifts from dirty cloths suspended in the wash water. The grease does not go back into the clothes.

Some chemicals make an emulsion thicken. The emulsion becomes a gel. A gel traps suspended particles. They do not settle out. A gel is a jellylike substance.

MIXING OIL AND WATER

What You Need (Materials)

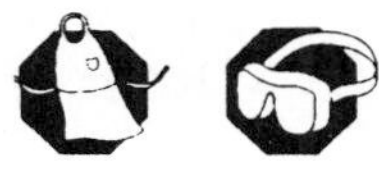

2 test tubes	oil
test tube rack	water

Figure A

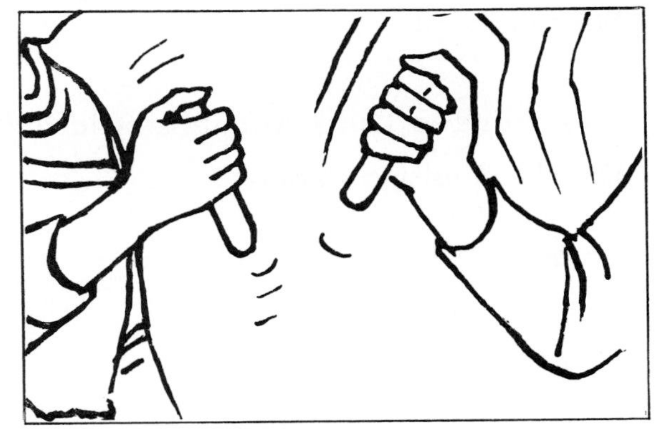

Figure B

How To Do The Experiment (Procedure)

1. Add the same amount of oil and water to each of the two test tubes.
2. Shake both at the same time.
3. Shake one gently.
4. Shake the other one very hard.
5. Put them back in the rack.

What You Learned (Observations)

1. Draw pictures of how the mixtures look right after shaking.

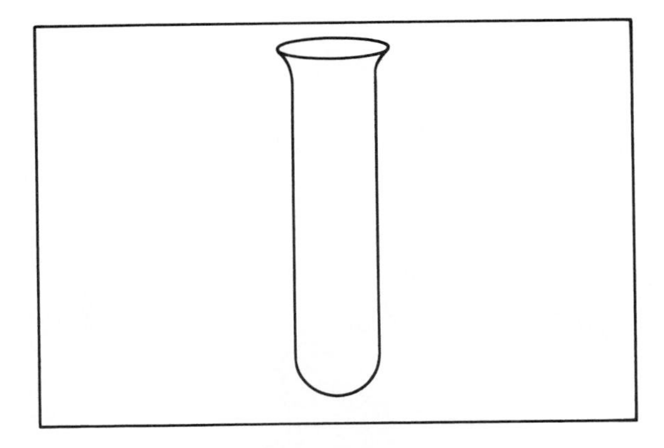

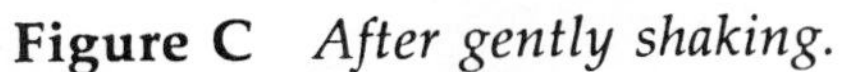

Figure C *After gently shaking.*

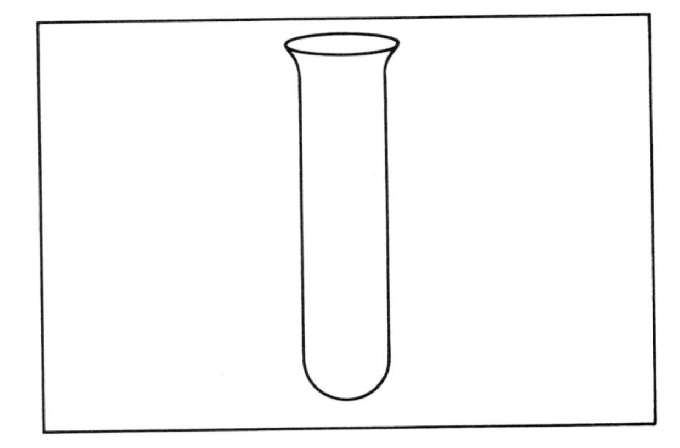

Figure D *After hard shaking.*

2. Fill in this chart. Put a check (✓) in the box that describes what happened.

	Made Size of Oil Drops Smaller	Left Larger Size Drops of Oil	Separated Slower	Separated Faster
Shaking Gently				
Shaking Hard				

Something to Think About (Conclusion)

Did the emulsion stabilize? ______________ Explain your answer. ______________

____

____

ADDING AN EMULSIFIER

What You Need (Materials)

2 test tubes
test tube rack
oil
water
detergent

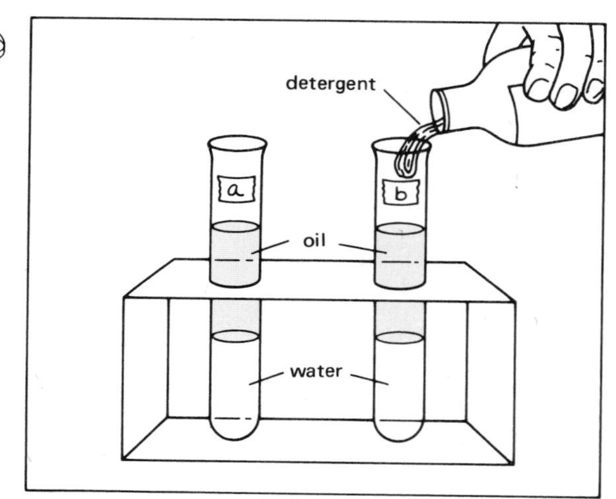

Figure E

How To Do The Experiment (Procedure)

1. Add equal amounts of oil and water to two test tubes. Label the test tubes a and b.
2. Let the test tubes stand until the parts separate.
3. Add some liquid detergent to test tube b.
4. Cover both test tubes and shake hard.
5. Let them stand for a few minutes. See what happens as they stand.

Figure F

What You Learned (Observations)

Look at both test tubes.

1. The parts in test tube ______________ separated quickly.
   a, b

2. The parts in test tube ______________ did not separate quickly.
   a, b

3. Detergent was added to test tube ______________ .
   a, b

4. The detergent ______________ the emulsion.
   stabilized, separated

Something To Think About (Conclusion)

Detergent is an example of ______________ .
a colloid, an emulsifying agent

ABOUT MILK

Milk is a common drink that is a suspension. It is mostly water with proteins, butterfat, sugar, vitamins, and minerals in it. But the parts of milk do not settle out because it has been **homogenized** [huh-MAHJ-uh-nyzed]. Fresh milk is a temporary emulsion that separates quickly into milk and cream. Fresh milk is homogenized in a machine that breaks down the cream into small particles. The small particles of cream stay permanently suspended in the milk.

Figure G *Milk that is not homogenized.*

tiny droplets
of fat

Figure H *Homogenized milk.*

FILL IN THE BLANK

Complete each statement using a term or terms from the list below. Write your answers in the spaces provided.

suspension	emulsion	homogenized
shake	gel	soap
detergents	emulsifying agents	adding special chemicals
stabilized	in layers	making the pieces smaller

1. Liquids that do not mix together make an ______________ .
2. An emulsion is a kind of ______________ .
3. Emulsions separate ______________ .
4. Any emulsion that does not separate is said to be ______________ .
5. We can stabilize emulsions by ______________________

   or ______________________ .
6. The parts of an emulsion become smaller when we ______________ it.
7. ______________ milk does not separate.
8. Chemicals that stabilize emulsions are called ______________ .
9. ______________ and ______________ are examples of emulsifiers.
10. A ______________ is a jellylike substance.

MATCHING

Match each term in Column A with its description in Column B. Write the correct letter in the space provided.

	Column A	Column B
________	1. homogenizing	a) stabilizes mayonnaise
________	2. egg yolk	b) jellylike substance
________	3. soaps and detergents	c) kind of suspension
________	4. emulsion	d) stabilizes milk
________	5. gel	e) chemical emulsifying agents

TRUE OR FALSE

In the space provided, write "true" if the sentence is true. Write "false" if the sentence is false.

_________ **1.** Emulsions separate in layers.

_________ **2.** We can keep emulsions from settling.

_________ **3.** Small particles settle fast.

_________ **4.** Hard shaking makes particles larger.

_________ **5.** Milk is homogenized with chemicals.

_________ **6.** Soaps and detergents are chemical emulsifying agents.

_________ **7.** Water is a gel.

_________ **8.** A gel is not as soft as a liquid and not as hard as a solid.

WORD SEARCH

The list on the left contains words that you have used in this Lesson. Find and circle each word where it appears in the box. The spellings may go in any direction: up, down, left, right, or diagonally.

EMULSIFY
EGG YOLK
STABILIZE
MILK
EMULSION
CLOUDY
CLEAR
COLLOID

L	E	S	L	I	E	D	O	W	A	R
D	M	I	L	L	I	E	I	Z	A	E
I	E	G	G	Y	O	L	K	E	L	Z
V	L	M	A	N	A	T	L	M	A	I
A	Y	D	U	O	L	C	F	E	K	L
D	E	M	U	L	S	I	O	N	D	I
F	T	L	I	M	S	N	Y	T	G	B
G	A	R	Y	K	L	I	M	O	U	A
D	I	O	L	L	O	C	F	L	D	T
C	L	L	A	D	N	Y	T	Y	T	S

REACHING OUT

Many of the foods we eat are processed. Things are done to the food. One process kills harmful germs.

1. You have learned that milk is homogenized. What else is done to store-bought milk that makes it different from "raw" milk?

____

Many chemicals are added to our food. These are food additives. Emulsifiers and stabilizers are food additives. Other additives change the color or flavor of food. Some keep the food fresher longer. Eating large amounts of these additives may be harmful to your health.

2. Look at the foods in the picture. Which of these foods do you think have food additives? _______________

____

____

3. Bring in some food labels from home. READ THE LABELS. List the food additives you find.

____

____

Figure I

What is a solution?

6

dissolve: go into solution
solute [SAHL-yoot]: substance that is dissolved in a solvent
solution: mixture in which one substance is evenly mixed with another substance
solvent: substance in which a solute dissolves

LESSON 6 What is a solution?

You can be a magician! Just add sugar to water and stir. Abracadabra! The sugar disappears!

But is the sugar really gone? No. The sugar and water just mix together. They mix completely. The sugar just seems to disappear.

Sugar and water together form a mixture. There are several kinds of mixtures. Sugar and water form a special kind of mixture. They form a liquid **solution.** There are many kinds of liquid solutions.

A liquid solution has two parts: a **solute** [SAHL-yoot] and a **solvent.** The solvent is always a liquid. The solute is what "disappears" in the solvent. The solute may be a solid, a gas, or another liquid.

A liquid solution is formed when the solute **dissolves.** The solute spreads out evenly throughout the solvent. The substance that dissolves is said to be soluble [SAHL-yoo-bul].

In the example of sugar and water, the water is the solvent. Water <u>dissolves</u> sugar. The sugar is the solute. Sugar is soluble in water.

There are many kinds of solvents. There are many kinds of solutes. There are many kinds of liquid solutions.

Remember, a mixture is a liquid solution <u>only</u> if the solute dissolves and spreads out evenly.

All the examples in Figures A, B, and C are liquid solutions.

Remember, there are three states of matter—solid, liquid, and gas.

Figure A

1. Name the states of matter of the substances in this liquid solution (Figure A).

   ____________ and ____________

2. The solute is the

   ________________ .
   solid, liquid

3. The solvent is the

   ________________ .
   solid, liquid

Figure B

4. Name the states of matter of the substances in this liquid solution (Figure B).

   ____________ and ____________ .

5. The solute is the

   ________________ .
   gas, liquid

6. The solvent is the

   ________________ .
   gas, liquid

Figure C

7. Name the states of matter of the substances in this liquid solution (Figure C).

   ____________ and ____________ .

NOTE: In solutions where all the parts are liquid, we usually do not name the solute and solvent.

COMPLETING SENTENCES

Choose the correct word or term for each statement. Write your choice in the spaces provided.

1. A liquid solution has at least one ______________ .
   solid, liquid, gas

2. The solute in a liquid solution ______________________________ .
   must be a gas, must be a solid, can be any state of matter

3. In solutions of liquids and solids or of liquids and gases, the solvent is always the ______________ .
   solid, liquid, gas

4. In solutions of all liquids, we usually ______________ name the solute and solvent.
   do, do not

WHICH ARE LIQUID SOLUTIONS?

Ten mixtures you know are listed below. Some are liquid solutions, some are not. Think about each mixture, then fill in the boxes.

	Mixture	Do the substances dissolve? (Write YES or NO.)	If the substances dissolved, name the solute (or solutes).	name the solvent.
1.	sugar water			
2.	muddy water			
3.	salty water			
4.	pebbles in water			
5.	instant coffee drink			
6.	orange juice			
7.	oil and water			
8.	instant tea drink			
9.	ocean water			
10.	vegetable soup			

FILL IN THE BLANK

Complete each statement using a term or terms from the list below. Write your answers in the spaces provided. Some words may be used more than once.

mixture	liquid solution	sugar water
two	liquid	solvent
solid	gas	solute
soluble	water	

1. Different things close together make up a ______________ .
2. A ______________ is a special kind of mixture.
3. An example of a liquid solution is ______________ .
4. A liquid solution has ______________ main parts.
5. One part of a liquid solution is always a ______________ .
6. The liquid part of a liquid solution is called the ______________ .
7. The other part of a liquid solution can be a ______________ , or a ______________ , or a ______________ .
8. The part of a liquid solution that mixes into the solvent is called the ______________ .
9. A solute that dissolves in a solvent is said to be ______________ .
10. Sugar is soluble in ______________ .

MATCHING

Match each term in Column A with its description in Column B. Write the correct letter in the space provided.

	Column A	Column B
________	1. mixture	a) means "able to dissolve"
________	2. solute	b) liquid part of a liquid solution
________	3. solvent	c) different things close together
________	4. liquid solution	d) a special kind of mixture
________	5. soluble	e) part of a solution that is dissolved

REACHING OUT

Be a detective! How can you tell if a mixture is a liquid solution? We will learn more in following Lessons. Meanwhile, see if you can figure out the clues.

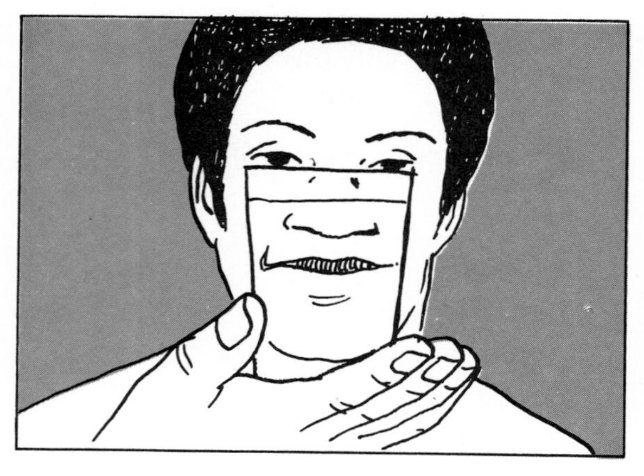

Figure D

- This is a mixture of sugar and water
- Sugar and water is a liquid solution

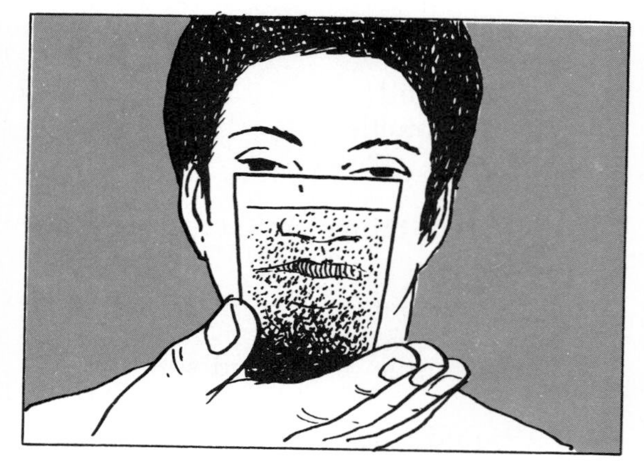

Figure E

- This is a mixture of muddy water.
- Muddy water is not a liquid solution.

Answer YES or NO to these questions.

		Muddy Water	Sugar Water
1.	Are the parts evenly mixed?		
2.	Can you see the separate parts?		
3.	Do particles fall to the bottom?		
4.	Can you see clearly through this mixture?		

How can you tell if a mixture is a liquid solution?

In your own words, list the clues.

____

____

____

____

____

What can we observe about liquid solutions?

7

immiscible [i-MIS-uh-bul]: not mixable
miscible [MIS-uh-bul]: mixable

LESSON 7 | What can we observe about liquid solutions?

Albert Einstein, Marie Curie, Louis Pasteur. . . . These are the names of some great scientists. What made them great? They looked around and saw things in a new way. They made new and important observations and conclusions.

Making observations is something that you can do, too. In this Lesson you will do some experiments with liquid mixtures. You will observe what happens to them.

Experiments and observations are part of the scientific method. The scientific method is a way of finding the answer to a question in several steps.

Scientists use scientific method to solve problems. You can use the scientific method, too. It is usually the most sensible way to solve any problem.

These are the steps to the scientific method:

(1) Start with a question.

(2) Get all the information about the question.

(3) Take a guess at what the best answer is.

(4) Plan an experiment to show if that is the answer.

(5) Do the experiment. Make careful observations.

(6) From your observations, decide if your guess was right or wrong. This is you conclusion. Sometimes you cannot make a conclusion. You have to do the experiment over and over!

The experiments in this Lesson have been planned for you. All you have to do is do them! Remember, do not just look—observe.

WHAT HAPPENS WHEN WE MIX A SOLID AND A LIQUID?

Part I Mixing Copper Sulfate and Water

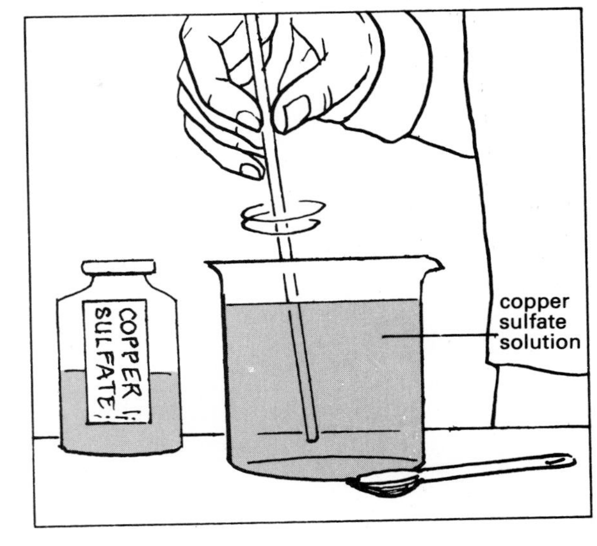

Figure A

Figure B

What You Need (Materials)
beaker
water
powered copper sulfate
scoop
stirrer

How To Do The Experiment (Procedure)

1. Pour some water into the beaker.
2. Add a small amount of powered copper sulfate to the water.
3. Stir well.

What Your Learned (Observations)

		YES	NO
1.	Does copper sulfate dissolve in water?		
2.	Does the mixture have a color?		
3.	Is the color the same all over?		

4. The parts of this mixture ______________ completely mixed.
   (are, are not)

5. This mixture ______________ a liquid solution.
   (is, is not)

6. The solvent is ______________ .
   (water, copper sulfate)

7. The solute is ______________ .
   (water, copper sulfate)

PART II Mixing Salt and Water

What You Need (Materials)

beaker
water
salt crystals (NaCl)
scoop
stirrer

Figure C

How To Do The Experiment (Procedure)

Add a small amount of salt to the water and stir well.

What You Learned (Observations)

		YES	NO
1.	Does the salt dissolve in water?		
2.	Does the mixture have a color?		
3.	Does the mixture look the same all over?		

4. The parts of this mixture ______________ (are, are not) completely mixed.

5. All the salt ______________ (does, does not) dissolve.

6. This mixture ______________ (is, is not) a solution.

7. Salt is a ______________ (solid, gas).

8. The salt is the ______________ (solute, solvent).

9. The water is the ______________ (solute, solvent).

Something To Think About (Conclusions)

1. A liquid solution ______________ (can, cannot) have a color.

2. In a liquid solution that has color, the color ______________ (is, is not) the same all over.

3. The color of a liquid solution is the same all over because the parts

_______________ mixed completely.
are, are not

4. Look at the solutions closely. You _______________ see the separate parts of a liquid solution.
can, cannot

WHAT HAPPENS WHEN WE MIX LIQUIDS?

When two or more liquids mix completely, we say they are **miscible** [MIS-uh-bul]. Let us test three liquids and see if they are miscible in water.

What You Need (Materials)

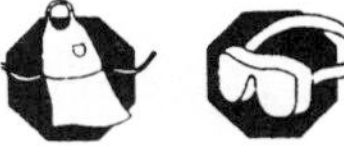

3 beakers　　vinegar　　3 stirring rods
alcohol　　oil

How To Do The Experiment (Procedure)

1. Fill 3 beakers half full with water.
2. Add a small amount of oil to one beaker.
3. Add a small amount of vinegar to another beaker. Stir, and let stand for two minutes.
4. Add a small amount of alcohol to the third beaker. Stir, and let stand for two minutes.

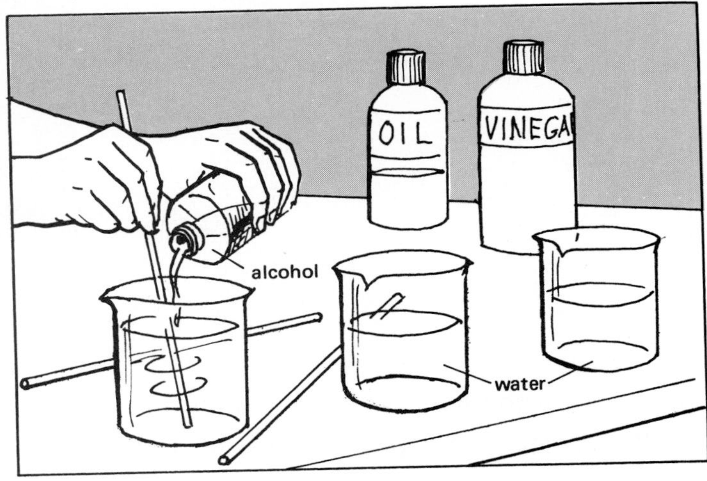

Figure D

What You Learned (Observations)

Write your observations in the chart below. Answer YES or NO.

		Alcohol and Water	Oil and Water	Vinegar and Water
1.	Do the liquids mix completely?			
2.	Are the liquids miscible?			
3.	Do they form a solution?			

Something to Think About (Conclusions)

1. Only certain liquids are ________________ in water.

2. Only certain liquids form a ______________________ with water.

WHAT HAPPENS WHEN WE MIX A GAS AND A LIQUID?

Let us see if a gas dissolves in a liquid. The gas we most often use is carbon dioxide. Carbon dioxide is one of the gases of the air. It also is a gas we exhale.

How can we tell when carbon dioxide is present? Easy! When it is in a solution of clear limewater, it turns milky.

What You Need (Materials)

beaker
drinking straw
limewater
plain tap water

Figure E

How To Do the Experiment (Procedure)

1. Pour some plain tap water into one of the beakers.

2. Using a straw, blow into the beaker.

3. The water ________________________________.
   stayed clear, turned cloudy

4. Pour a small amount of limewater into the beaker and stir.

5. The water ________________________________.
   stayed clear, turned cloudy

6. Pour some tap water into the second beaker.

7. Pour a small amount of limewater into the second beaker and stir.

8. The water ________________________________.
   stayed clear, turned cloudy

9. Using a straw, blow into the beaker

10. The water ________________________________.
    stayed clear, turned cloudy

What You Learned (Observations)

1. The solute is the ______________ .
   carbon dioxide, water

2. The solvent is the ______________ .
   carbon dioxide, water

Something To Think About (Conclusions)

1. This experiment shows that carbon dioxide ______________ dissolve in water.
   did, did not

2. Carbon dioxide ______________ soluble in water.
   is, is not

THE SCIENTIFIC METHOD

Use these words to describe the scientific method.

conclusion	experiment	question
observations	information	guess

1. Start with a ______________ .
2. Get all the ______________ about the question.
3. Take a ______________ at the answer.
4. Plan an ______________ .
5. Make careful ______________ of what you see.
6. Finally, from your observations make your ______________ .

FILL IN THE BLANK

Complete each statement using a term or terms from the list below. Write your answers in the spaces provided. Some words may be used more than once.

colorless	immiscible	salt water
solute	liquid	solvent
oil and water	miscible	water and alcohol
states of matter	mixture	

1. A liquid solution is a special kind of ______________ .
2. The main parts of a liquid solution are the ______________ and ______________ .
3. The solute of a liquid solution may be any of the three ______________________.
4. The solvent of a liquid solution is always a ______________ .
5. Some liquid solutions have a color. Others are ______________ .
6. Two examples of liquid solutions are ______________ and ______________ .
7. Liquids that form a liquid solution are said to be ______________ .
8. An example of miscible liquids are ______________ .
9. Take a guess! The word that means not miscible is ______________ .
10. An example of immiscible liquids is ______________ .

MATCHING

Match each term in Column A with its description in Column B. Write the correct letter in the space provided.

	Column A		Column B
______	1. miscible	a)	means "able to dissolve"
______	2. soluble	b)	mixable
______	3. solute	c)	special kind of mixture
______	4. solvent	d)	liquid part of a liquid solution
______	5. liquid solution	e)	can be a solid, liquid or gas

Does water dissolve everything?

8

tincture: [TINK-chur]: solution of a substance in alcohol

LESSON 8 | Does water dissolve everything?

You get a stain on your clothes. What do you reach for first? Probably water. You know from experience that water is a good solvent. It dissolves many things.

Water is the most used solvent. It is so popular, it is called the "universal" [yoo-nuh-VUR-suhl] solvent.

Water may be a good solvent, but it does not dissolve everything. No one solvent dissolves everything! Special solvents are needed to dissolve certain substances.

You have probably put iodine on a cut. Well, here is a surprise. Iodine is a solid, not a liquid. It has been dissolved in alcohol. The solution is called "tincture [TINK-chur] of iodine." The iodine would not dissolve in water. A solution that has alcohol for the solvent is called a **tincture.**

If you put salt in water, the salt dissolves. But if you put salt in alcohol, it does not dissolve.

So, alcohol dissolves iodine, but not salt. Water dissolves salt, but not iodine. Solvents are selective. This means they dissolve certain things but not others.

USES FOR SOLVENTS

Figure A

Figure B

Figure C

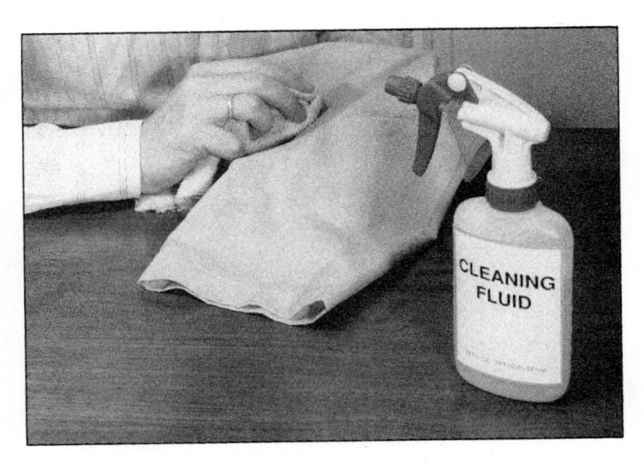

Figure D

TRUE OR FALSE

In the space provided, write "true" if the sentence is true. Write "false" if the sentence is false.

________ 1. Water is the most popular solvent.

________ 2. Water dissolves everything.

________ 3. Alcohol is called the universal solvent.

________ 4. Water dissolves salt.

________ 5. Iodine is a liquid.

________ 6. All solutions are tinctures.

________ 7. Only solutions with alcohol are tinctures.

FILL IN THE BLANK

Complete each statement using a term or terms from the list below. Write your answers in the spaces provided. Some words may be used more than once.

tincture	sugar	solid
water	universal	alcohol
iodine	salt	

1. The most popular solvent is ______________ .
2. Water is such a popular solvent that it is called the ______________ solvent.
3. Two solids that water can dissolve are ______________ and ______________ .
4. A solid that water cannot dissolve is ______________ .
5. Iodine is a ______________ .
6. Iodine does not dissolve in ______________ .
7. Iodine does dissolve in ______________ .
8. A solution that uses alcohol as a solvent is called a ______________ .

REACHING OUT

Turpentine or paint thinner often is used to clean paint brushes. When would turpentine or paint thinner not be used to clean paint brushes? Explain.

____

____

____

____

What are the properties of solutions?

9

homogeneous [hoh-muh-JEE-nee-us]: uniform; the same all the way through
properties [PROP-ur-tees]: characteristics used to describe a substance
transparent [trans-PER-unt]: material that transmits light easily

LESSON 9 | What are the properties of solutions?

What happens when you add salt to a jar of water and stir? The salt disappears. You have made a liquid solution. Does the same thing happen when you add sand to water? No. The sand settles to the bottom of the jar.

How can we tell if a mixture is a solution or not? We can tell by its **properties** [PROP-ur-tees]. Properties tell us how a kind of matter looks and acts.

These are the properties of liquid solutions:

(1) The parts dissolve and become the size of molecules.
(2) Liquid solutions are **homogenous** [hoh-muh-JEE-nee-us].
(3) Liquid solutions are **transparent** [trans-PER-unt].
(4) Liquid solutions do not settle out.

MOLECULE SIZE You know that matter is made up of tiny atoms. Most matter is made up of groups of atoms called molecules. In a liquid solution, the particles of solute dissolve. They break up until they are the size of molecules.

HOMOGENOUS Homogeneous means evenly mixed—the same all though. Because the particles are the size of molecules they weigh very little. They move around and spread out evenly.

TRANSPARENT You can see clearly through something that is transparent. Glass is transparent. So are liquid solutions. The molecules that make them up are tiny. They do not block out light. Light passes right through.

THE PARTS NEVER SETTLE OUT Something that settles out drops to the bottom of its container. The parts of a liquid solution never separate. They never settle out no matter how long they sit. That is because the molecules are light. They keep bouncing around. This also keeps the solution homogenous.

MORE ABOUT SOLUTIONS

Figure A

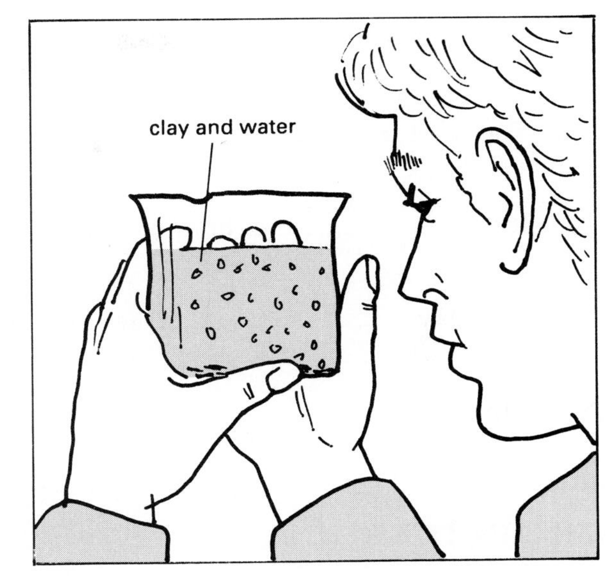

Figure B

Look at Figure A.

1. Can you see the sugar particles? ______________ (yes, no)

2. The sugar ______________ (did, did not) dissolve.

3. The sugar is now ______________ (the size of molecules, much larger than the size of molecules).

4. Can the boy see through the sugar water? ______________ (yes, no)

5. The sugar water is ______________ (cloudy, transparent).

6. The mixture ______________ (is, is not) evenly mixed.

7. It ______________ (is, is not) homogenous.

8. The sugar ______________ (is, is not) settling.

9. Sugar water ______________ (is, is not) a liquid solution.

Look at Figure B.

1. Can you see the clay particles? ______________ (yes, no)

2. The clay ______________ (did, did not) dissolve.

3. The clay particles are ______________ (the size of molecules, much larger than the size of molecules).

4. Can the boy see clearly through the mixture? _______________
yes, no

5. The clay water is _______________ .
cloudy, transparent

6. The mixture _______________ evenly mixed.
is, is not

7. It _______________ homogenous.
is, is not

8. The clay _______________ settling out.
is, is not

9. Clay water _______________ a liquid solution.
is, is not

FILL IN THE BLANK

Complete each statement using a term or terms from the list below. Write your answers in the spaces provided. Some words may be used more than once.

liquid solutions	moving around	light
drop	is not	clay water
molecules	transparent	small in size

1. When we can look clearly through something we say it is _______________ .
2. _______________ are transparent.
3. _______________ is not transparent.
4. Clay water _______________ a liquid solution.
5. The parts of a liquid solution are the size of _______________ .
6. The molecules of a liquid solution do not block _______________ .
7. To "settle out" means to _______________ .
8. The parts of _______________ do not settle out.
9. Liquid solutions do not settle out because the parts are too

   _______________ .
10. The molecules in liquid solutions are always _______________ .

WHICH IS HOMOGENOUS?

The dots stand for copper sulfate molecules. The liquid is water.

Figure C **Figure D** **Figure E**

1. Which figure shows a homogenous mixture? ______________

2. a) The mixtures in Figures ________ and ________ are not liquid solutions.

   b) They are not liquid solutions because they ______________ (are, are not) homogenous.

3. a) The mixtures that are not liquid solutions ______________ (could, could not) become liquid solutions.

   b) They would be solutions if all the ______________ (solute, solvent) dissolved, and spread out evenly.

4. Think about this: What would you do to make the mixtures that are not homogenous, become homogenous fast?

____

____

MATCHING

Match each term in Column A with its description in Column B. Write the correct letter in the space provided.

	Column A		Column B
________	1. molecule	a)	evenly mixed
________	2. homogenous	b)	drop
________	3. settle out	c)	tiny part of matter
________	4. properties	d)	clear
________	5. transparent	e)	things that help us identify matter

TRUE OR FALSE

In the space provided, write "true" if the sentence is true. Write "false" if the sentence is false.

_______ 1. Anything we can see through clearly is transparent.

_______ 2. Every mixture is homogenous.

_______ 3. Sand becomes the size of molecules when it is in water.

_______ 4. Liquid solutions are transparent.

_______ 5. Muddy water is transparent.

_______ 6. Muddy water settles out.

_______ 7. The parts of liquids are the size of molecules.

_______ 8. Salt water is a liquid solution.

_______ 9. Liquid solutions settle out.

_______ 10. The molecules of solutions are always moving around.

REACHING OUT

Transparent, translucent, and opaque are three words that have to do with light. Give a definition of each word in the spaces below. (You may use a dictionary.) Next to each definition give an example of each.

Transparent ______________________

____

____

Translucent ______________________

____

____

Opaque ______________________

____

____

Figure F

How can the strength of a solution be changed?

10

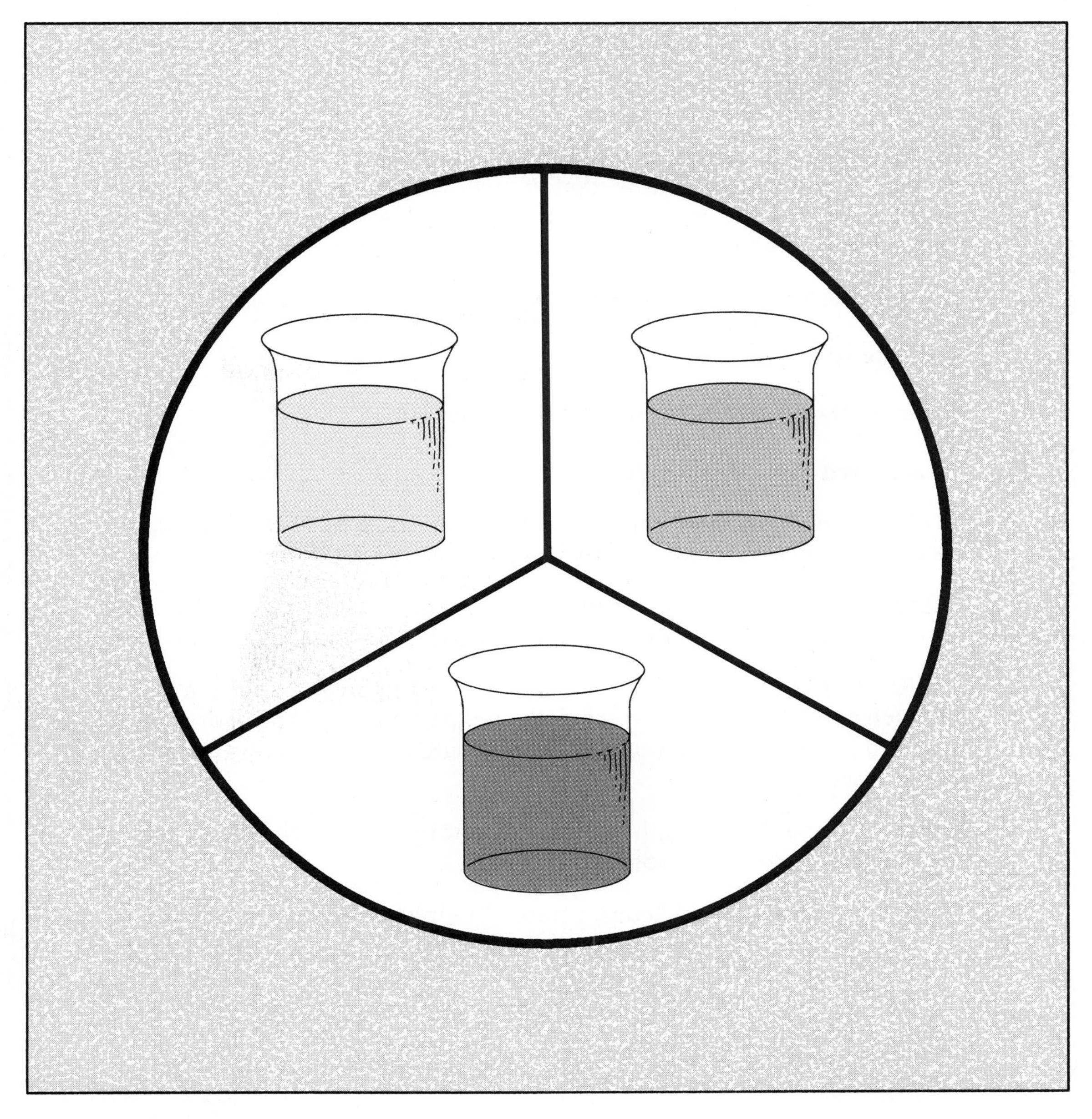

concentrated [KAHN-sun-trayt-ed] **solution:** strong solution
dilute: [di-LEWT] **solution**: weak solution
saturated [SACH-uh-rayt-id] **solution:** solution containing all the solute it can hold at a given temperature

LESSON 10 | How can the strength of a solution be changed?

Some people like strong coffee. Others like it weak. What makes coffee strong or weak? You know that if you add more coffee it becomes stronger. If you add more water, it becomes weaker. A cup of coffee is like any liquid solution. It comes in many strengths.

We use these terms to describe how strong a liquid solution is:

- **dilute** [di-LEWT] **solution**
- **concentrated** [KAHN-sun-trayt-ed] **solution**
- **saturated** [SACH-uh-rayt-id] **solution:**

DILUTE A dilute solution is a weak solution. It has very little solute dissolved in the solvent.

CONCENTRATED A concentrated solution is a strong solution. It has more solute dissolved in it than a dilute solution has.

SATURATED A saturated solution is an extra-strong solution. It has so much solute that no more can dissolve. If we tried to add more solute, it would just drop to the bottom. More solute would dissolve only if the mixture were heated.

We can change the strength of a liquid solution. We do this by changing the amount of solute or solvent.

Terms like "dilute" and "concentrated" help us compare solutions.

HOW STRONG IS THE SOLUTION?

A cup of instant tea is a liquid solution.

Two different cups of tea are shown in Figures A and B. Study them. Then fill in the blanks.

Figure A

Figure B

1. The instant tea ________________ (does, does not) dissolve.

2. A cup of tea ________________ (is, is not) a liquid solution.

3. In both solutions the solute is the ________________ (tea, water).

4. In both solutions the solvent is the ________________ (tea, water).

5. The cup of tea in Figure A has a ________________ (lot, little bit) of solute compared to the solution in Figure B.

6. The solution in Figure A is ________________ (stronger, weaker) than the solution in Figure B.

7. The solution in Figure A is more ________________ (concentrated, dilute) than the solution in Figure B.

8. The solution in Figure B ________________ (has, does not have) the same kind of solute and solvent as Figure A.

9. The amount of solvent is ______________ in both solutions.
the same, different

10. The amount of solute is ______________ in both solutions.
the same, different

11. There is ______________ solute in Figure B.
more, less

12. This solution is ______________ than the solution in Figure A.
stronger, weaker

13. This solution is ______________ than the solution in Figure A.
more concentrated, more dilute

COLOR AND STRENGTH OF SOLUTIONS

Sometimes, color tells us about the strength of a solution. The darker the color, the stronger the solution. The lighter the color, the weaker the solution. We can use color only when we compare similar solutions.

Of Figures A and B on page 61:

1. which figure shows a darker solution? ______________

2. which figure shows a lighter solution? ______________

3. which solution is stronger? ______________

4. which solution is weaker? ______________

5. Why can color help us compare the strengths of these mixtures? ______________

____

____

COMPARING CONCENTRATIONS

Figure C | Figure D | Figure E

Look at Figures C, D, and E. Each test tube contains a liquid solution. The kind of solute and solvent is the same in each.

1. Which is the strongest? ____________

2. Which is the weakest? ____________

3. Which has the most solute? ____________

4. Which has the least solute? ____________

5. Which one is closest to being saturated? ____________

A SATURATED SOLUTION

Figure F

Figure G

The beakers in Figures F and G contain liquid solutions. The kind of solute and solvent is the same in each. Some additional solute has been added to each. Both mixtures have been stirred well.

1. Which is not saturated? ____________

2. Which one is saturated? ____________

3. How do you know? ____

____

4. Take a guess! How can we make the extra solute dissolve? ____________

____

____

FILL IN THE BLANK

Complete each statement using a term or terms from the list below. Write your answers in the spaces provided.

lighter	strengths	saturated
very little	raising the temperature	a lot
color	concentrated	darker
dilute		

1. Liquid solutions come in different ______________ .
2. A weak solution is called ______________ .
3. A strong solution is called ______________ .
4. Dilute solutions have ______________ solute.
5. Concentrated solutions have ______________ of solute.
6. A solution that can dissolve no more solute is called ______________ .
7. We can make a saturated solution dissolve more solute by

   ______________________ .
8. Sometimes we can use ______________ to compare the strengths of solutions.
9. In comparing strength by color, the ______________ the color, the stronger the solution.
10. In comparing strength by color, the ______________ the color, the weaker the solution.

MATCHING

Match each term in Column A with its description in Column B. Write the correct letter in the space provided.

	Column A	Column B
______	1. dilute solution	a) extra solvent does not dissolve
______	2. concentrated solution	b) more solute dissolves
______	3. saturated solution	c) strong solution
______	4. raising the temperature	d) sometimes used to compare solution strengths
______	5. color	e) weak solution

WORD SEARCH

The list on the left contains words that you have used in this Lesson. Find and circle each word where it appears in the box. The spellings may go in any direction: up, down, right, or diagonally.

MOLECULE
SOLUTE
SOLVENT
DILUTE
SOLUBLE
GAS
DISSOLVE
SOLID
MIXTURE
LIQUID
SATURATE

Y	M	I	X	T	U	R	E	D	C
R	S	O	L	U	T	E	U	I	L
A	O	S	L	I	D	V	N	S	O
G	L	T	N	E	V	L	O	S	S
E	U	D	M	P	C	S	L	O	E
T	B	R	I	T	X	U	E	L	L
A	L	U	L	L	S	D	L	V	B
R	E	S	E	O	U	A	M	E	I
U	D	O	A	B	E	T	A	P	C
T	D	L	Q	G	I	Z	E	R	S
A	U	I	L	E	S	L	I	E	I
S	Q	D	I	U	Q	I	L	A	M

REACHING OUT

Why can color be used only for comparing similar solutions and not for comparing different solutions?

____

____

____

____

SCIENCE *EXTRA*

Environmental Chemist

Are you interested in how chemistry affects your daily life or the environment around you? Do you enjoy solving problems? If you do, you might enjoy a career as an environmental chemist.

Environmental chemists deal mainly with pollution. They analyze air, soil, or water samples to find out if pollutants are present. They determine how these pollutants combine and react with one another. Environmental chemists also try and find out what these reactions mean to people and the climate. Some environmental chemists suggest solutions to pollution problems. Other environmental chemists are responsible for the cleanup of polluted areas.

Environmental chemists were used in 1989 when an oil tanker hit a reef in Alaska. About 11 million gallons of oil leaked into the water. Fish and other marine animals that lived in the area died because of the pollution. Marshes and wildlife refuges were destroyed as well.

Environmental chemists were sent to Alaska to survey the damage caused by the oil spill. They suggested ways of cleaning up the spill to help restore the area. Some environmental chemists helped in the clean up.

Environmental chemists have a variety of career options open to them. For example, some environmental chemists may work in industry or for the federal government. Other environmental chemists find work doing research or teaching.

To be an environmental chemist, you need a college degree in chemistry or environmental science. Some people go on and get advanced degrees in environmental science.

How can solutes be made to dissolve faster?

11

LESSON 11 | How can solutes be made to dissolve faster?

What do you do after you add sugar to a drink? You stir. But why? You stir because you know that mixing makes sugar dissolve faster.

Stirring makes any solute dissolve faster.

Now here is another question. Which will dissolve faster, a lump of sugar or small grains of sugar? You know from experience that the smaller the pieces the faster it will dissolve.

Now let us tackle another question. Which will dissolve sugar faster, cold water or hot water?

From your experience you know that solutes dissolve faster in hot water.

There are three ways to make solids dissolve faster:

- Stir the mixture.
- Break the solute into smaller pieces.
- Heat the mixture.

Doing any of these things makes a solute dissolve faster. Doing two or all three dissolves the solute much faster.

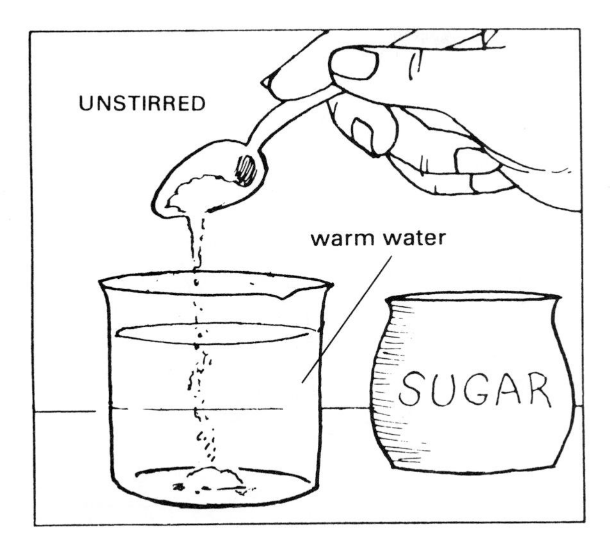

Figure A

Pour some granulated sugar into a glass of warm water. Do not stir.

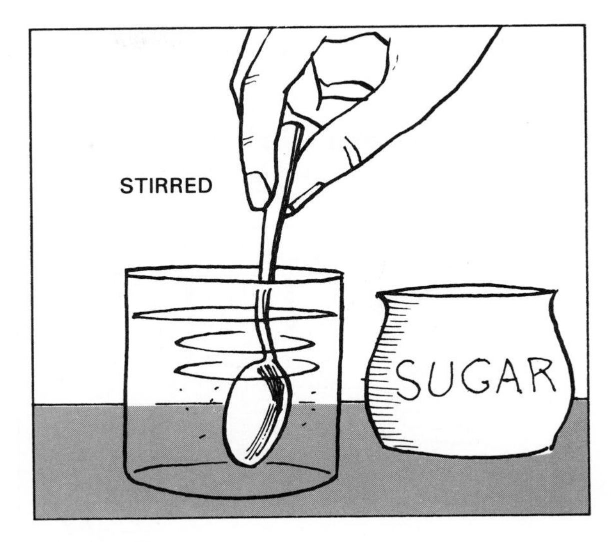

Figure B

Pour the same amount of granulated sugar into another glass of warm water. Then stir.

Notice how fast the sugar dissolves in each glass.

1. Stirring makes a solute dissolve ________________ .
   faster, slower

Figure C

Pour some granulated sugar into cold water. Do not stir.

Figure D

Pour the same amount of granulated sugar into hot water. Do not stir.

Notice how fast the sugar dissolves in each beaker.

2. A solute dissolves faster in a ________________ solvent.
hot, cold

3. A solute dissolves slower in a ________________ solvent.
hot, cold

4. Heat makes a solute dissolve ________________ .
slower, faster

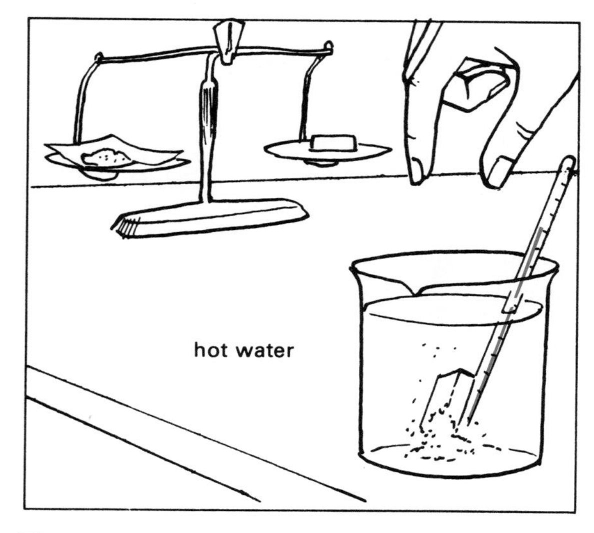

Figure E

Put a lump of sugar into a glass of hot water. Do not stir.

Figure F

Pour one teaspoon of granulated sugar into a glass of hot water. Do not stir.

Notice how fast the sugar dissolves.

5. Big pieces dissolve ________________ than small pieces.
slower, faster

6. Small pieces dissolve ________________ than large pieces.
slower, faster

7. Crushing makes solutes dissolve ________________ .
slower, faster

NOW TRY THESE

1. The two main parts of any liquid solution are the ________________ and the ________________ .

2. The liquid part of a solution is called the ________________ .

3. The part of a solution that dissolves into the liquid is called the

_______________ .

4. Three ways we can make a solid solute dissolve faster are:

_________________________ , _________________________ ,

and _________________________ .

COMPLETE THE CHART

Complete the chart by placing a check under the correct response.

	If you . . .	then the solute dissolves faster.	slower.
1.	make the pieces larger,		
2.	make the pieces smaller,		
3.	stir,		
4.	do not stir,		
5.	heat the solvent,		
6.	do not heat the solvent.		

TRUE OR FALSE

In the space provided, write "true" if the sentence is true. Write "false" if the sentence is false.

________ **1.** Stirring moves things around.

________ **2.** Crushing makes things larger.

________ **3.** Heat lowers temperature.

________ **4.** Stirring makes solutes dissolve faster.

________ **5.** Small pieces of solute dissolve slower than big pieces.

________ **6.** Heat makes solutes dissolve faster.

WORD SCRAMBLE

Below are several scrambled words you have used in this Lesson. Unscramble the words and write your answers in the spaces provided.

1. NEAHIGT ______________________
2. TARSEF ______________________
3. RIST ______________________
4. SCURH ______________________
5. SLEDSIVO ______________________

REACHING OUT

You put a piece of glass in water. It does not dissolve. You then crush it into tiny pieces.

Will the pieces dissolve? ______________ Explain. ______________________

____

____

____

____

What is boiling point?

12

boiling point: temperature at which a liquid changes to a gas
evaporate [i-VAP-uh-rayt]: change from a liquid to a gas
water vapor: water in the gas state

LESSON 12 | What is boiling point?

Brrr. . . . Did you ever take a bath in cold water? Few of us have. Almost everyone bathes in hot water.

Heated water is very important. We use it in many ways—for cleaning—for cooking. Some homes are heated by hot water. Doctors use boiling water to kill germs on instruments. The steam from boiling water runs ocean liners and navy ships—even submarines.

What happens when water is heated? Follow what happens step-by-step. As you read, check with Figure A.

(1) When water is heated, its temperature rises (a).

(2) At first you see tiny bubbles. They form on the sides and bottom of the container (b). They are air bubbles. They had been dissolved in the water. The heat is forcing them out of the water.

(3) The temperature keeps rising as you keep heating the water.

When the temperature reaches 100° C (212° F), you see large bubbles. They rise quickly from the bottom of the water (c). These bubbles tell us that the water is boiling.

The bubbles are **water vapor**—water that has changed to a gas. The water **evaporates** [i-VAP-uh-rayts]—the liquid changes to a gas. The gas escapes into the air. First the vapor is hot, invisible steam. It quickly changes to a fine cloud of tiny drops of liquid water. It is this fine cloud that most people call steam (d).

The temperature at which a liquid changes to a gas is called its **boiling point.** The boiling point for plain water is 100° C.

(4) The water keeps boiling. But its temperature does not rise any higher. It stays at 100° C.

Unless something special happens, plain water does not get hotter than 100° C. Adding more heat just makes the water evaporate more rapidly. Soon all the water boils away.

WHAT HAPPENS WHEN WATER IS HEATED?

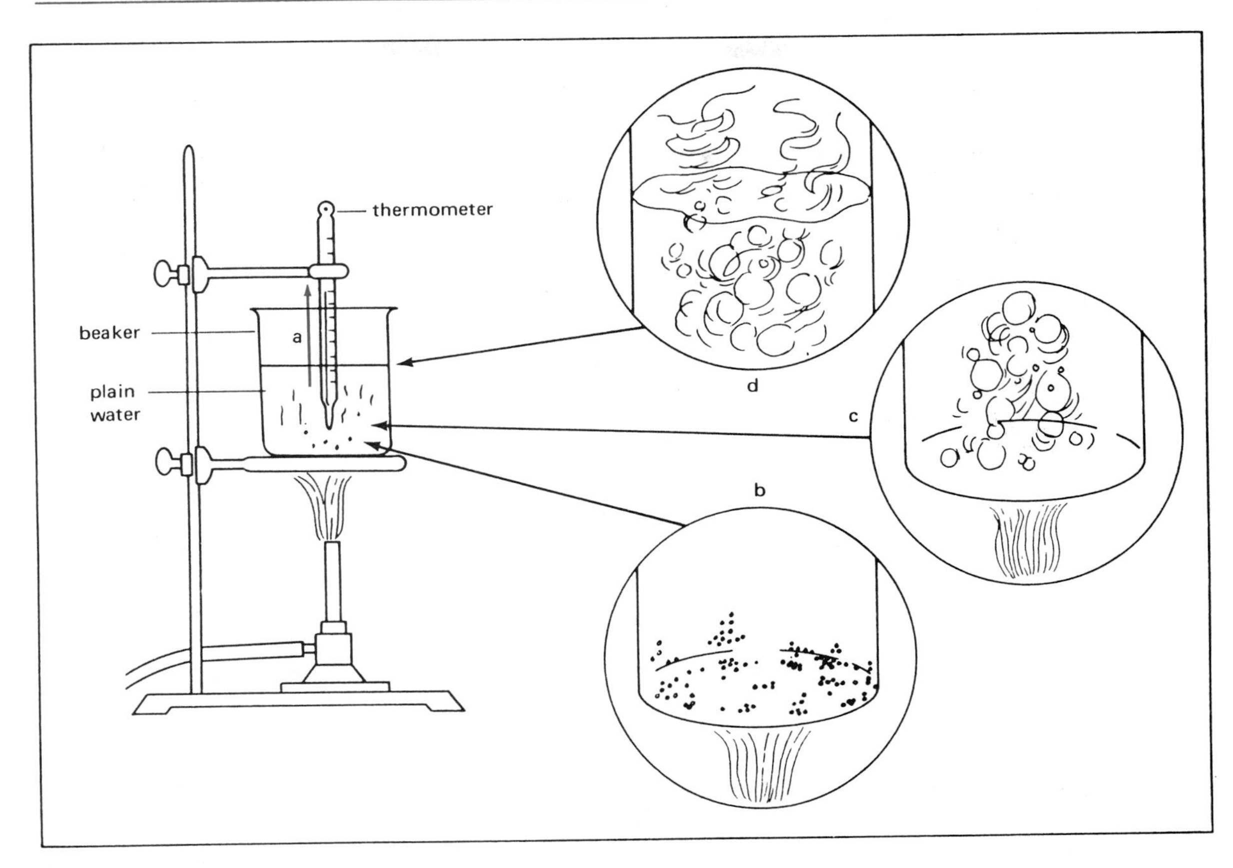

Figure A

What You Need (Materials)

ring stand and clamp
beaker
plain water
thermometer
Bunsen burner
watch or clock

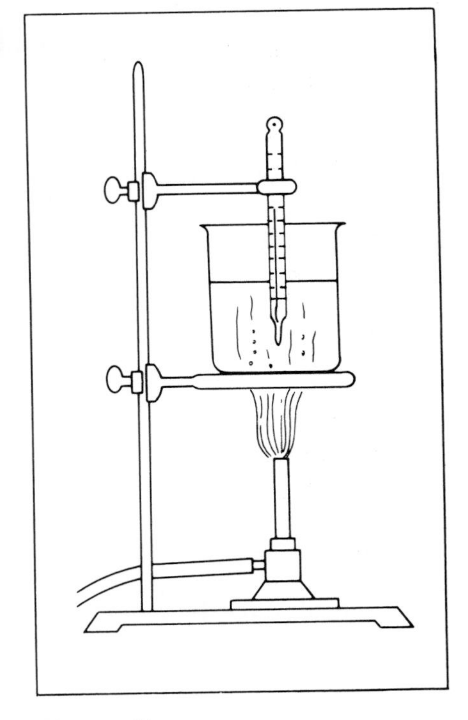

Figure B

How To Do The Experiment (Procedure)

1. Set up the materials as shown in Figure B. Do not light Bunsen burner yet.
2. Take the temperature of the water before it is heated. Write it down on the chart on the next page.
3. Light the flame under the beaker. Take the water temperature again at the times listed on the chart. Write down the temperature each time.
4. During all this time watch the water. See what happens. (Notice the water level too!)

What You Learned (Observations)

1. Fill in the following chart.

		Temperature
	Before Heating	
During Heating	1 minute	
	2 minutes	
	3 minutes	
	at boiling	
	1 minute after boiling	
	2 minutes after boiling	
	3 minutes after boiling	

1. What happened to the temperature of water when it was heated? ______________

____

2. At what temperature did the water boil? ______________° C.

3. How did you know that the water was boiling? (What did you see?) ______________

____

4. You kept the water boiling for three minutes. Did the water get any hotter while boiling? ______________

5. What does this prove about the temperature of boiling water? ______________

____

6. What happened to the water level as the water kept boiling? ______________

____

Something to Think About (Conclusions)

1. What would happen to the water if you were to keep the water boiling?

____

2. Think back to when the water was just starting to heat up.

   a) Did you notice small bubbles? ______________

   b) What were these bubbles? ______________

____

TRY THIS AT HOME

1. Fill a glass with cold tap water.
2. Let it stand overnight.
3. Describe what you see the next day.

   ____

   ____

4. Draw a picture of what you see on Figure C.

Figure C

FILL IN THE BLANK

Complete each statement using a term or terms from the list below. Write your answers in the spaces provided.

rises	liquid solution	large bubbles
water vapor	100° C	dissolved air
liquid	thermometer	gas
does not		

1. A ______________________ measures temperature.
2. When matter is heated, its temperature ______________ .
3. The water you drink is in the ______________ state.
4. Tiny bubbles form soon after water is heated. They are bubbles of

   ______________________ .

5. The tiny air bubbles prove that water is a ______________________ .
6. When water boils, ______________________ quickly rise to the surface.
7. The large bubbles are ______________________ .
8. Water vapor is water in the ______________ state.
9. Tap water boils when its temperature reaches ______________ .
10. If we keep on boiling water, its temperature ______________ rise any higher.

TRUE OR FALSE

In the space provided, write "true" if the sentence is true. Write "false" if the sentence is false.

________ **1.** Tap water has air dissolved in it.

________ **2.** Water can hold more dissolved air when it is heated.

________ **3.** A thermometer tells us how long something is heated.

________ **4.** Heat raises temperature.

________ **5.** Tiny bubbles on the side of a pot of water means that the water is boiling.

________ **6.** If water boils for a long time, its temperature rises above 100° C.

________ **7.** Boiled water has gases dissolved in it.

________ **8.** Water is always evaporating.

________ **9.** Water evaporates fastest when it boils.

________ **10.** You can see water vapor.

________ **11.** There is water in the air.

REACHING OUT

What happens to the water vapor that goes into the air?

____

____

____

____

____

____

Figure D

What is a freezing point? 13

freezing point: temperature at which a liquid changes to a solid

LESSON 13 What is a freezing point?

Can you walk on water? Sure you can! Just step onto ice. Ice is water. It is in the solid state, but it is still water—H_2O.

Water can be a liquid, or a solid, or a gas. It all depends on the temperature.

You have already learned that water changes to a gas when it takes in enough heat.

What happens when water loses heat?

When water loses heat, its temperature drops. It becomes colder and colder. When the water loses enough heat, it changes to a solid called ice. We say it freezes.

Plain water freezes at 0° C. This is its **freezing point.** Other liquids freeze at different temperatures. When water temperature falls to 0° C, it stays liquid for a short time. Then it changes to a solid—a bit at a time.

The temperature of ice keeps dropping as the air temperature around it becomes colder. This means that some ice is colder than other ice.

DOES ICE HAVE HEAT?

Yes, ice does have heat!

All matter has heat. Some matter just has more heat than other matter.

Figure A

Heat moves from one place to another.

Heat that is lost in one place goes somewhere else.

As matter becomes warmer, it gains heat.

As matter becomes cooler, it loses heat.

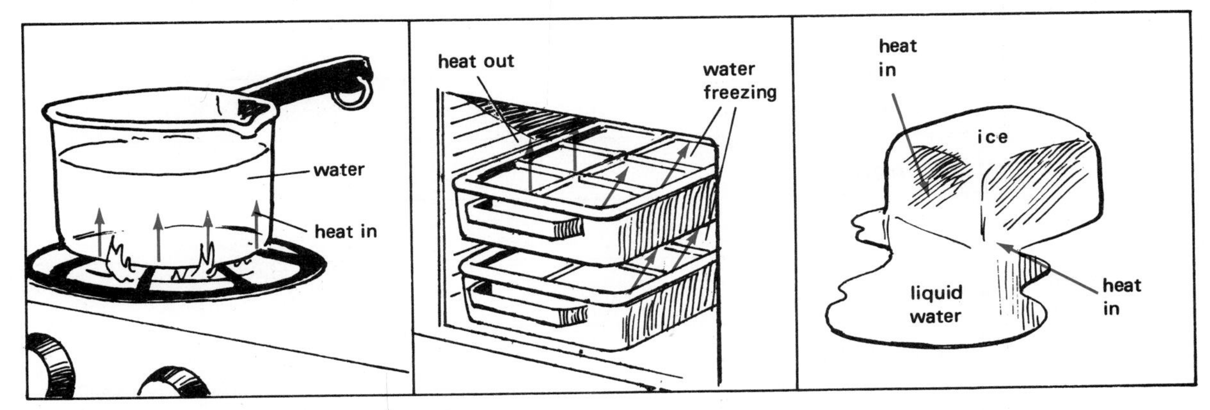

Figure B *When a liquid becomes warmer, it gains heat.*

Figure C *When liquid water freezes, it loses heat.*

Figure D *When ice changes to liquid water, it gains heat.*

WHAT HAPPENS WHEN WATER FREEZES?

This experiment can be done at home.

Purpose: To find the freezing temperature of water.

What You Need (Materials)

small plastic cup half full with water
thermometer that goes below 0° C (32° F)
refrigerator freezer

What To Do (Procedure)

1. Check the water's temperature before placing it in the freezer. Write it down on the chart below.
2. Keep the thermometer in the cup. Place the cup into your freezer.
3. Check the temperature every 15 minutes. Write down the temperature each time. Do this until all the water changes to ice.
4. Check the temperature in the morning. Write it down on the chart.

Figure E

	Time	**Temperature**	**State of the Water (Liquid, Part Liquid/ Part Solid, All Solid)**
Outside the freezer	**Start**		**All Liquid**
Inside the freezer	15 min.		
	30 min.		
	45 min.		
	1 hour		
	1 hr., 15 min.		
	1 hr., 30 min.		
	2 hrs.		
	2 hrs., 30 min.		
	next morning		

What You Learned (Observations)

1. While the water was in the freezer, its temperature went ______________ (up, down).

2. When the temperature just reached 0° C (32° F), the water ______________ (was still a liquid, changed to ice).

3. The water then changed to ice ______________ (all at once, a little at a time).

4. What was the temperature of the water as it was changing to ice? ______________

5. What was the temperature when all the water changed to ice? ______________

6. a) What was the temperature of the ice in the morning? ______________

   ____

   b) What does this show about the temperature of ice? ______________

   ____

   ____

Something To Think About (Conclusion)

When water freezes, it ______________ (takes in, gives off) heat.

TRUE OR FALSE

In the space provided, write "true" if the sentence is true. Write "false" if the sentence is false.

______ 1. Everything has heat.

______ 2. Heat can be taken in and given off.

______ 3. Something that takes in heat gets colder.

______ 4. Heat that is given off by one thing is taken in by something else.

______ 5. Water is always a liquid.

______ 6. Water can be a gas.

______ 7. Water can be a solid.

______ 8. Water vapor is water in the solid state.

______ 9. Water freezes at 0° C.

______ 10. 0° C and 32° F stand for the same temperature.

CROSSWORD PUZZLE

Use the clues to complete the crossword puzzle.

Clues

Across

1. temperature at which a liquid changes to a solid
6. material that transmits light easily
7. go into solution
9. substance that is dissolved in a solvent
10. mixture in which one substance is evenly mixed with another substance
11. water in the gas state

Down

2. change from a liquid to a gas
3. characteristics used to describe a substance
4. solution of a substance in alcohol
5. temperature at which a liquid changes to a gas
7. weak solution
8. solution containing all the solute it can hold at a given temperature
9. substance in which a solute dissolves

SOLUTIONS

How can solutes change the freezing and boiling point of water?

14

LESSON 14 How can solutes change the freezing and boiling point of water?

Most automobile engines are cooled with water. In the winter, it is very cold in most parts of the country. If the water freezes, it can ruin the engine.

Car owners add antifreeze to the car's cooling system. This prevents the water from changing to ice. The same antifreeze also protects the engine from boiling over in the hot summer.

How does antifreeze work? Antifreeze acts like a solute in a solution. Putting antifreeze in the water of the cooling system raises the boiling point of the water. It also lowers the freezing point of the water.

Some dissolved solutes change the boiling and freezing point of water. Some dissolved solutes make it harder for water to freeze and boil.

This means that the water needs more cold to freeze and more heat to boil, The water freezes at a temperature lower than 0° C. It boils at a temperature higher than 100° C.

Adding more solutes lowers the freezing temperature and raises the boiling temperature even more. BUT THERE IS A LIMIT. After a certain amount of solute has been added, no more changes take place.

SOLUTES CHANGE BOILING AND FREEZING POINTS

Certain solutes

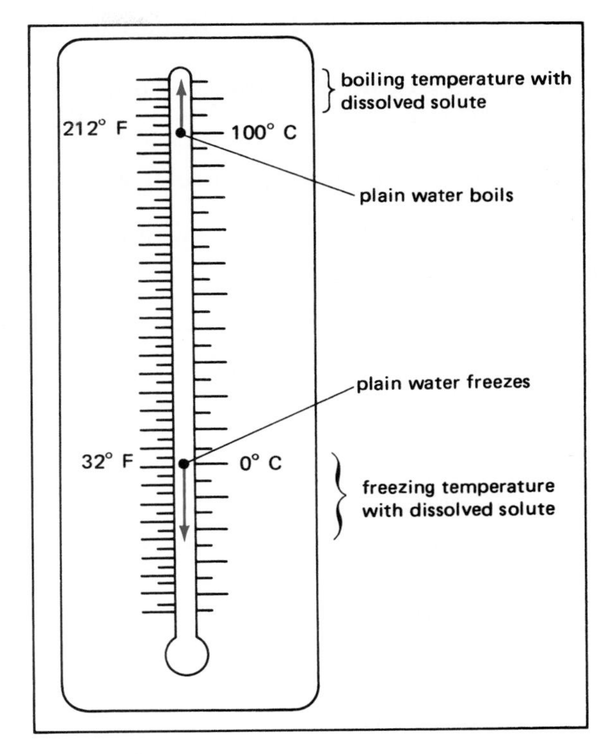

raise the boiling temperature,
and
lower the freezing temperature of water.

Figure A

HOW DOES SALT CHANGE THE BOILING OF WATER?

What You Need (Materials)

2 beakers	water
2 ring stands and clamps	salt
2 thermometers	2 Bunsen burners

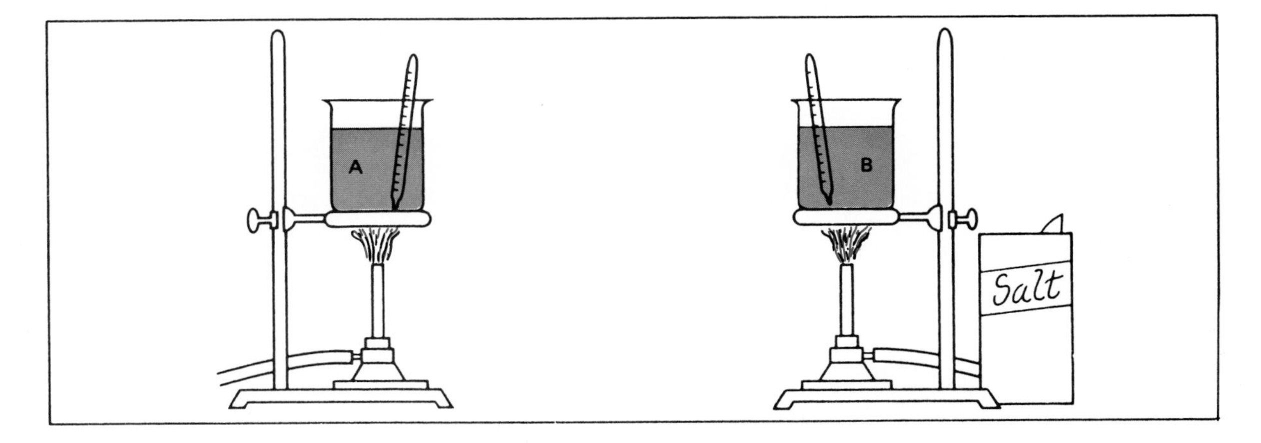

Figure B

How To Do The Experiment (Procedure)

1. Fill both beakers half full with water. Label one beaker A and the other B.
2. Stir a teaspoon of salt into beaker B only.
3. Set the beakers on the ringstands and place the thermometers in the water. Light the burners.
4. Observe the temperature at which the water boils in each beaker.
5. Put your observations on the chart below.

	Boiling Point
PLAIN TAP WATER	
SALT WATER	

What You Learned (Observations)

1. The tap water boiled at ______________° C.
2. The salt water boiled at ______________° C.
3. The salt water boiled at a ______________ (higher, lower) temperature than the tap water.
4. Water boils at a ______________ (higher, lower) temperature when a solute is added.

Something To Think About (Conclusions)

1. Is salt water a solution? ______________
2. Which part is the solute? ______________
3. Which part is the solvent? ______________
4. Do all dissolved solutes raise the boiling point of water? ______________

TRY THIS AT HOME

Get two small plastic containers of the same size. Fill them half full with tap water.

Dissolve a tablespoon of salt into one of the containers.

Label this container S.

Place the containers in your freezer.

Check them every half hour.

Which one freezes first?

____

____

____

Figure E

1. The plain tap water froze ________________ (faster, slower) than the salt water.

2. The salt water froze ________________ (faster, slower) than the plain tap water.

3. Dissolved salt water ________________ (easier, more difficult) to freeze.

TRUE OR FALSE

In the space provided, write "true" if the sentence is true. Write "false" if the sentence is false.

________ **1.** Adding solutes to water makes the water more difficult to freeze.

________ **2.** Adding solutes to water lowers the freezing point.

________ **3.** Adding solutes to water raises the freezing point.

________ **4.** Adding solutes to water makes it easier to boil.

________ **5.** Adding solutes to water raises the boiling point.

________ **6.** Adding solutes to water lowers the boiling points.

________ **7.** Antifreeze acts like a solute.

________ **8.** The freezing point of plain water is higher than the freezing point of salt water.

________ **9.** The boiling point of salt water is 100° C.

________ **10.** The freezing point of plain water is 0° C.

REACHING OUT

Why do you think people put rock salt on icy sidewalks?

____

____

____

____

____

____

____

____

Figure F

How can the parts of a solution be separated?

15

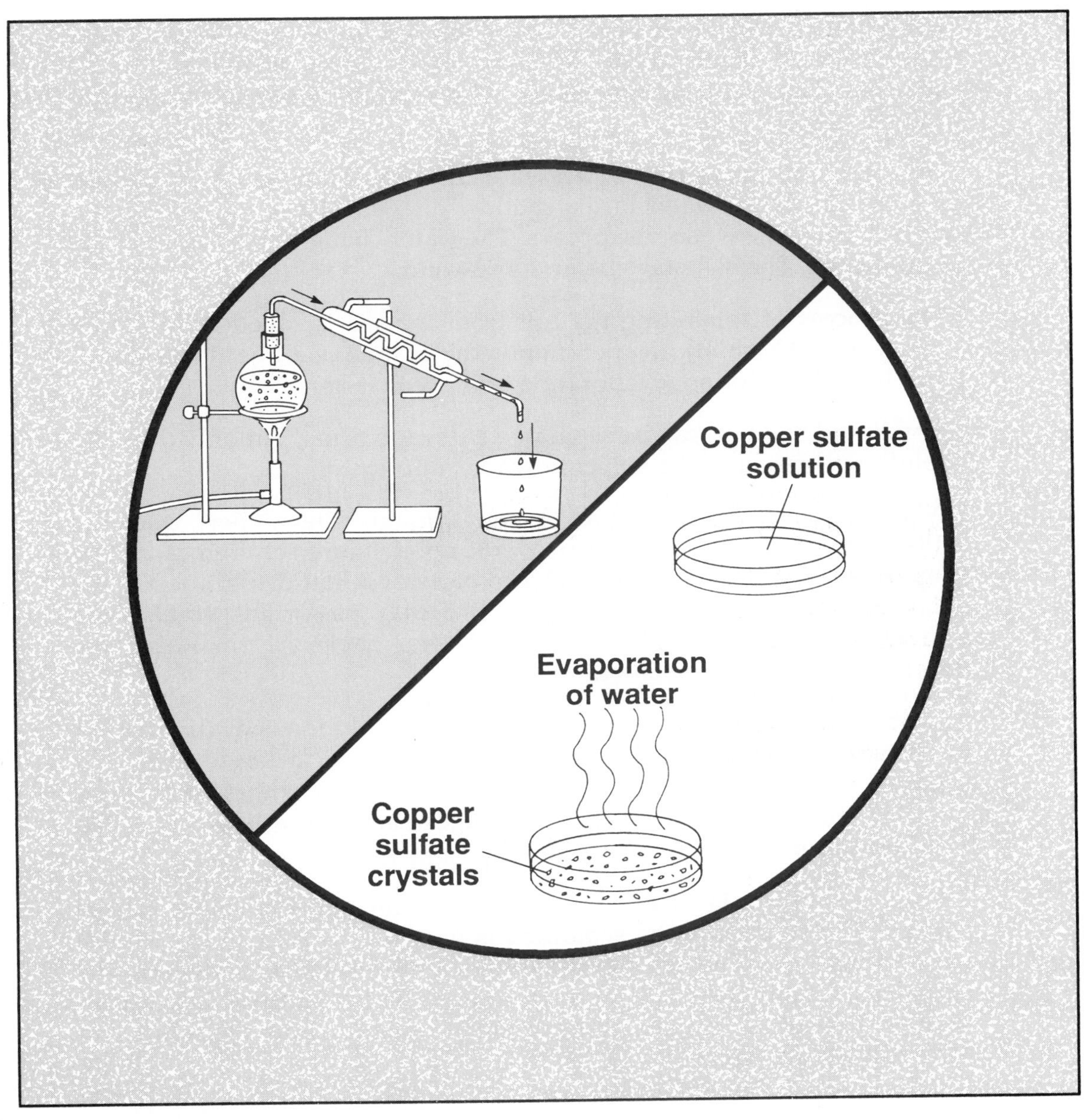

condensation [kahn-dun-SAY-shun]: change of a gas to a liquid
distillationn [dis-tuh-LAY-shun]: process of evaporating a liquid and then condensing the gas back into a liquid
evaporation [i-vap-uh-RAY-shun]: change of a liquid to a gas at the surface of the liquid

LESSON 15 | How can the parts of a solution be separated?

Everybody knows that ocean water tastes salty. Ocean water tastes salty because there is salt dissolved in it. Ocean water is a liquid solution. The water is the solvent. The salt is one of the solutes dissolved in it.

How can you prove that ocean water contains dissolved salt? Simple! Place some ocean water into a shallow dish and let it stand for a few days. Slowly the water disappears. The water changes to a gas and goes into the air. The salt stays behind as a solid.

The process of separating the salt from the sea is called **evaporation** [i-vap-uh-RAY-shun]. Evaporation is the change of a liquid to a gas. If you heat the solution, evaporation will occur faster.

Any liquid solution can be separated by evaporation, but only the solid solute will remain. The solvent escapes into the air.

What happens if you pass salt water through filter paper? Does the filter paper trap the salt? The answer is no! A liquid solution cannot be separated by filtering. Why not? The parts of a liquid solution are the size of molecules. They are so tiny that they pass right through the holes of the filter paper. The holes in filter paper are small, but the molecules are much smaller.

Another method of separating a solute from a solution is by **distillation** [dis-tuh-LAY-shun]. In the process of distillation, a liquid is heated until it evaporates. The gas is then cooled until it changes back into a liquid. The process by which gas changes back into a liquid is called **condensation** [kahn-dun-SAY-shun].

When a solution is distilled, both the solvent and the solute can be saved. The solution to be separated is heated. The solvent evaporates and forms a gas. The gas moves through a tube called a condenser. The condenser cools the gas back to a liquid. The liquid drips into a container. The solute remains in the original container. Both the solute and the solvent are saved.

EVAPORATION

See what happens when salt water evaporates. Figure A shows you the experiment.

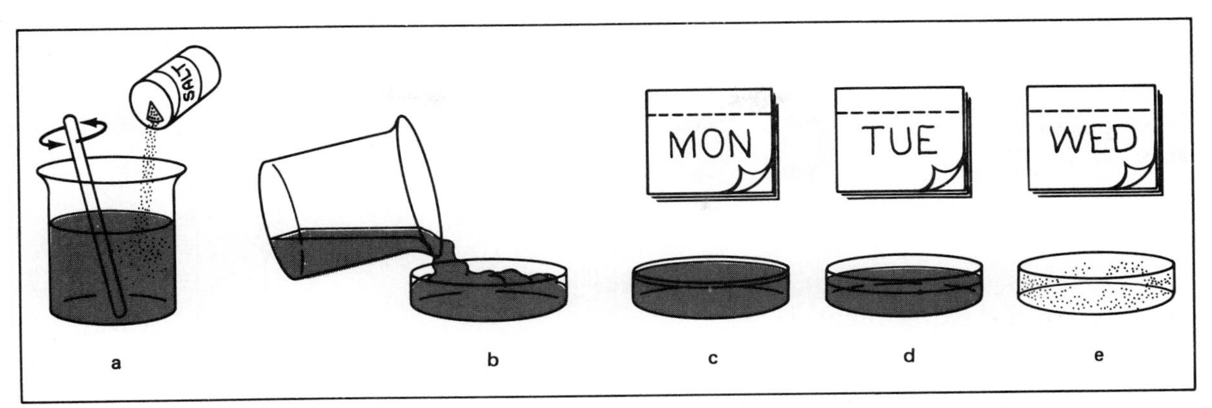

Figure A

1. Did the salt and water form a solution? ______________

2. In salt water, **a)** the solute is the ______________ (salt, water).

   b) the solvent is the ______________ (salt, water).

3. Did the water from your dish evaporate? ______________

4. Was anything left behind? ______________

5. The ______________ (solute, solvent) was left behind in its ______________ (solid, liquid) form.

6. Name the solid that was left behind. ______________

7. Name the liquid that went into the air. ______________

8. When salt water is left in an open dish, the ______________ (salt, water) evaporates.

9. In evaporation, a ______________ (gas, liquid) changes to a ______________ (gas, liquid).

10. When a solution evaporates, we get back ______________ (only the solute, only the solvent, both the solute and solvent).

CAN FILTERING SEPARATE A SOLUTE FROM A SOLVENT?

What You Need (Materials)

2 beakers
copper sulfate solution
funnel
filter paper
ring stand

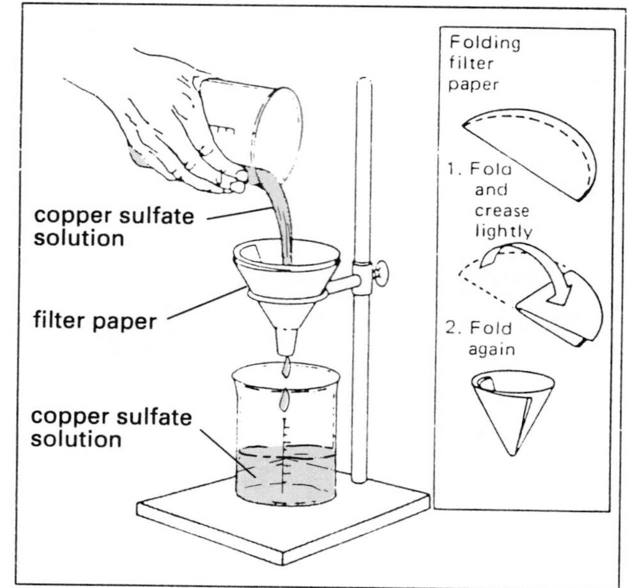

Figure B

How To Do the Experiment (Procedure)

1. Set up the ring stand with the funnel in place.
2. Fold the filter paper as shown in Figure B and place in the funnel.
3. Put the beaker under the funnel.
4. Pour the copper sulfate solution through the filter paper.

What You Learned (Observations)

1. The solute ______________ left behind in the filter paper.
   (was, was not)
2. The solvent ______________ left behind in the filter paper.
   (was, was not)
3. The solute particles are ______________ than the holes in the filter paper.
   (larger, smaller)
4. The solvent particles are ______________ than the holes in the filter paper.
   (larger smaller)

Something To Think About (Conclusions)

1. A liquid solution ______________ be separated by filtering.
   (can, cannot)
2. The parts of a liquid solution are ______________________________.
   (the size of molecules, larger than the size of molecules)
3. Filter paper holes are ______________________________.
   (the size of molecules, larger than the size of molecules)

A DISTILLATION UNIT

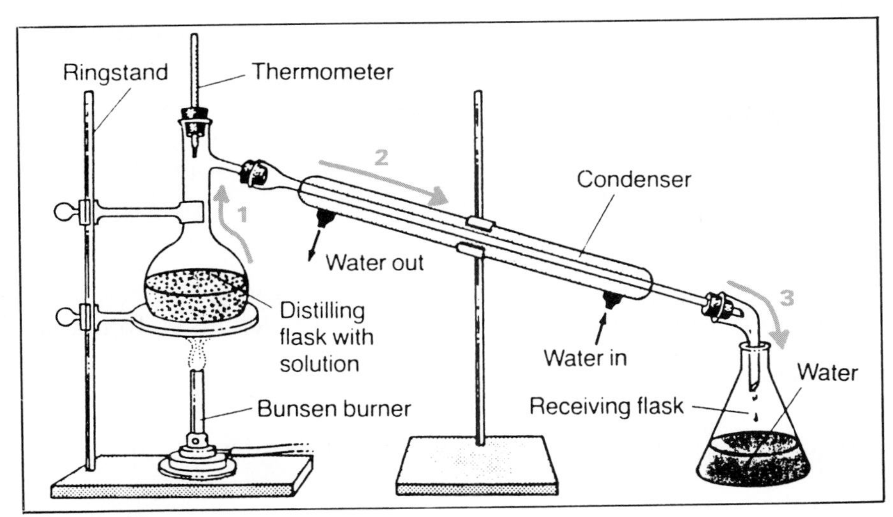

Figure C

How a Distillation Unit Works:

1. A solution is heated in a flask and the water turns to steam. Solids or liquids that have not reached their evaporation point remain the flask.
2. The steam enters the condenser and is cooled. As it cools, it changes back to a liquid.
3. The condensed liquid comes out of the condenser and enters the receiving flask.

WHAT HAPPENS IN DISTILLATION?

Check with Figure C as you read.

1. The liquid solution is boiled in the boiler. The solvent evaporates. The solvent changes from a ______________ to a gas (water vapor).
2. The vapor moves out of the boiler. It goes into the inner tube of the cooling section.
3. The cold water in the outer tube cools the vapor. This makes the vapor condense. The vapor changes from a ______________ back to a liquid.
4. The liquid drips into a container. It is pure. It has been distilled. It has no solute dissolved in it.
5. What happens to the solute? The solid solute stays behind in the boiler. It is now dried up. It is in solid form.

FILL IN THE BLANK

Complete each statement using a term or terms from the list below. Write your answers in the spaces provided. Some words may be used more than once.

filtering	distillation	drop
gas	larger	solvent
solute	heated	liquid
distilled		

1. When a liquid solution evaporates, only the ______________ changes to a gas.
2. Evaporation happens faster when a solution is ______________ .
3. Filter paper holes are ______________ than the size of molecules.
4. Liquid solutions cannot be separated by ______________ .
5. Evaporation saves only the ______________ of a liquid solution.
6. ______________ gets back both the solute and the solvent from a liquid solution.
7. In evaporation a ______________ changes to a ______________ .
8. In condensation a ______________ changes to a ______________ .
9. Condensation takes place when there is a ______________ in temperature.
10. ______________ water has no solutes in it.

WORD SCRAMBLE

Below are several scrambled words you have used in this Lesson. Unscramble the words and write your answers in the spaces provided.

1. NEDTASCONION ______________
2. TILLSIDNATION ______________
3. NIOTRAPAEOV ______________
4. LOTSUE ______________
5. THEA ______________

ACIDS AND BASES

What are acids?

16

acid: substance that reacts with metals to release hydrogen
indicator [IN-duh-kayt-ur]: substance that changes color in acids and bases

LESSON 16 | What are acids?

The sour taste of the lemon juice tells us that it is an **acid**. Acids are special kinds of chemicals. They are common in everyday life. Some are helpful, others are harmful. There are some that are weak. Others are strong. Many acids are dangerous to touch or taste. You should never touch or taste an unknown acid.

Besides the sour taste that acids have, there are other tests for identifying them. Certain chemicals change color when acids are added.

Chemicals that change color are called **indicators** [IN-duh-kayt-urz].

An example of an indicator is a litmus paper. Litmus paper comes in two colors, red and blue.

Acids turn blue litmus paper red. Acids do not change the color of red litmus paper.

When acids mixed with metals a chemical reaction takes place. Hydrogen gas is given off from this reaction.

TESTING FOR AN ACID

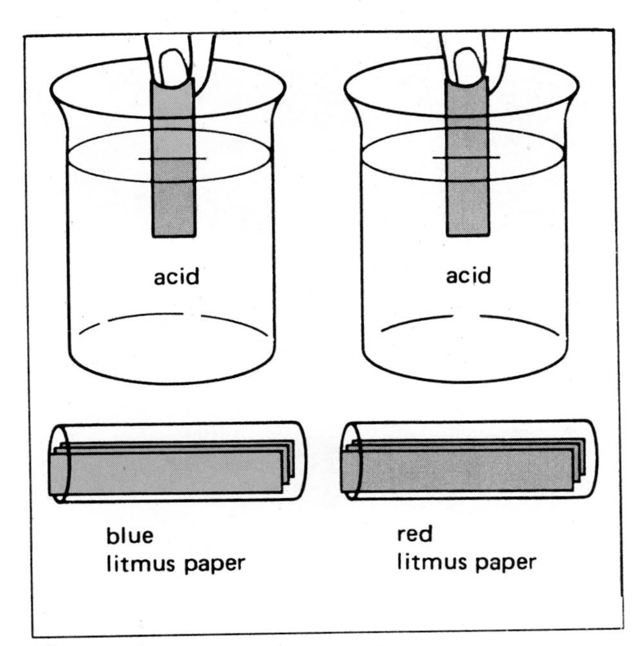

Figure A

Acids turn blue litmus paper red.

Does the red litmus paper change color with acids? ______________

SOME COMMON ACIDS

The chart lists some common acids and their chemical formulas. It shows you what all acids have in common. All acids contain the element hydrogen (H+).

	ACID	CHEMICAL FORMULA	USES
1.	Acetic acid	$HC_2H_3O_2$	vinegar
2.	Boric acid	H_3BO_3	eyewash
3.	Carbonic acid	H_2CO_3	club soda
4.	Citric acid	$H_3C_6H_5O_7$	citrus fruits
5.	Hydrochloric acid	HCl	aids digestion
6.	Nitric acid	HNO_3	fertilizers
7.	Sulfuric acid	H_2SO_4	plastics

Figure B *Citrus fruits have an acid called citric acid.*

Figure C *Vinegar is acetic acid.*

Figure D *Sulfuric acid is used in car batteries.*

Figure E *Hydrochloric acid produced in the stomach helps in digestion.*

FILL IN THE BLANK

Complete each statement using a term or terms from the list below. Write your answers in the spaces provided. Some words may be used more than once.

never	acid	blue
vinegar	hydrochloric acid	red
litmus paper	hydrogen	citric
dangerous		

1. Lemons contain ______________ acid.
2. ______________ is a kind of indicator.
3. Acids turn ______________ litmus paper red.
4. ______________ litmus paper does not change color in acids.
5. When acids wear away metals, ______________ is given off.
6. Acetic acid is found in household ______________.
7. Your stomach produces______________________________.
8. All acids contain the element ______________.
9. Some acids are ______________ to touch or taste.
10. You should ______________ touch or taste an ______________.

TRUE OR FALSE

In the space provided, write "true" if the sentence is true. Write "false" if the sentence is false.

________ 1. Litmus paper is an indicator.

________ 2. Acids turn red litmus paper blue.

________ 3. Acids contain hydrogen.

________ 4. Acids wear away metals.

________ 5. Oxygen is given off when acids wear down metals.

WORD SCRAMBLE

Below are several scrambled words you have used in this lesson. Unscramble the words and write your answers in the spaces provided.

1. CADI ____________________
2. TRINOCDIA ____________________
3. SSTTE ____________________
4. SITLUM ____________________
5. RUSO ____________________

REACHING OUT

Figure F

Sometimes when rain falls, it mixes with pollution particles in the air. An acid is formed.

Why might this be harmful? ____________________

What are bases?

17

base: substance formed when metals react with water

phenolphthalein [fee-nohl-THAL-een]: an indicator that turns a deep pink color when a base is added

LESSON 17 | What are bases?

Bases are a group of chemicals that have certain properties. Their properties are different from the properties of acids. Often they act opposite to the ways that acids act.

However, like acids, bases may be of different strengths. Some are very weak. Some are very strong. Some bases are dangerous to touch or taste. You should never touch or taste an unknown base.

Let us see how bases act with tests that we use to identify chemicals.

Bases have a bitter taste.

If you touch a harmless base it will feel slippery. Acids do not have any special feel.

Bases act the opposite way from acids with indicators. Bases turn red litmus paper blue. They do not change blue litmus paper.

There is another indicator that helps us to identify bases. It is called **phenolphthalein** [fee-nohl-THAL-een). This solution is clear in acids. But phenolphthalein turns deep pink in bases.

Unlike acids, bases do not wear away metals.

TESTING FOR A BASE

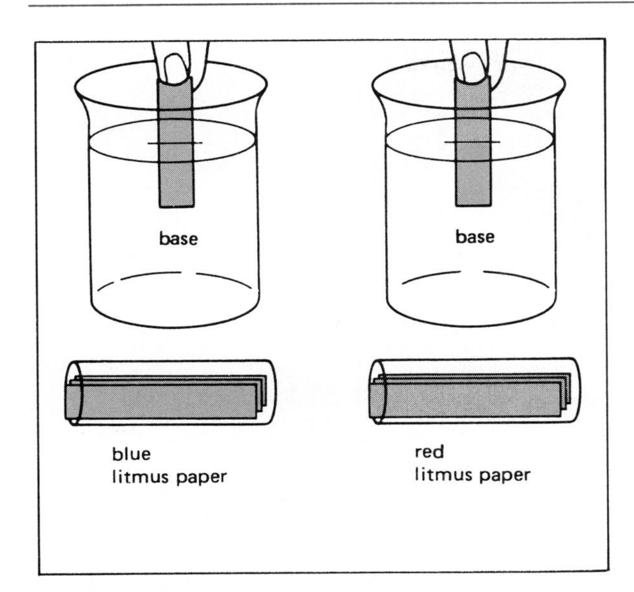

Figure A

Bases turn red litmus paper blue. Blue litmus paper does not change color.

What happens to blue litmus paper in acids? ______________________

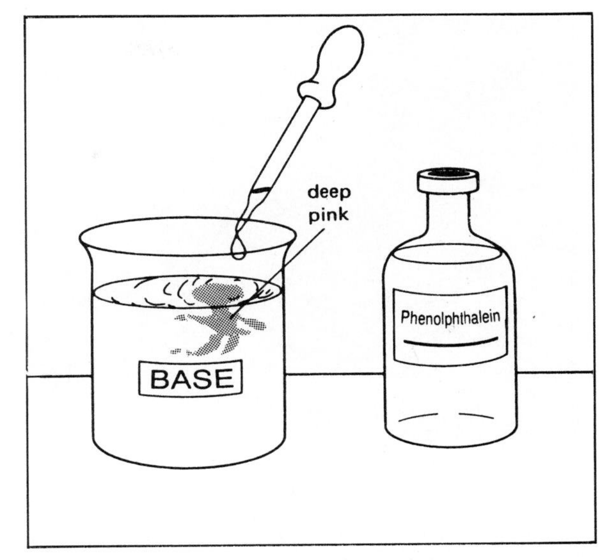

Figure B

Phenolphthalein turns deep pink in bases.

What happens to phenolphthalein in acids?

____

SOME COMMON BASES

The chart lists some common bases and their chemical formulas It shows you what all bases have in common. All bases contain special groups of oxygen and hydrogen atoms called hydroxides (OH^-).

	BASE	CHEMICAL FORMULA	USES
1.	Potassium hydroxide	KOH	soap
2.	Magnesium hydroxide	$Mg(OH)_2$	milk of magnesia
3.	Calcium hydroxide	$Ca(OH)_2$	mortar
4.	Ammonium hydroxide	NH_4OH	ammonia
5.	Sodium hydroxide	NaOH	soap

Figure C *Ammonia is used in cleaning products.*

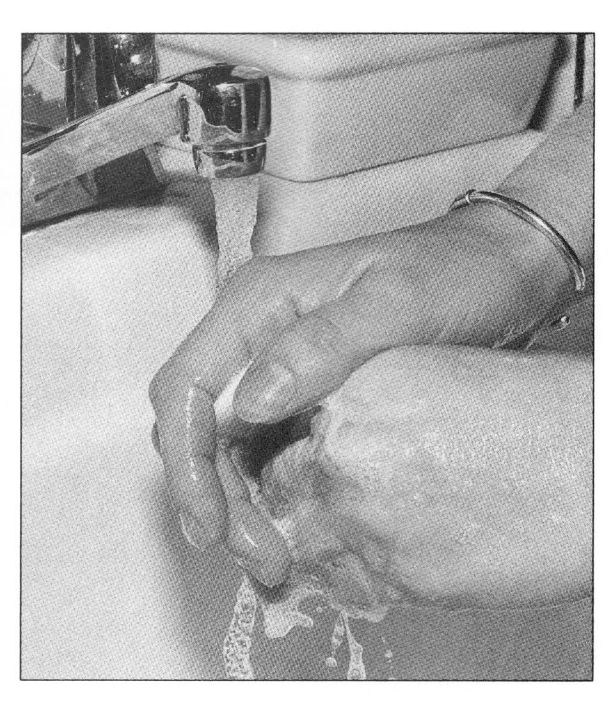

Figure D *Soap contains a base called lye.*

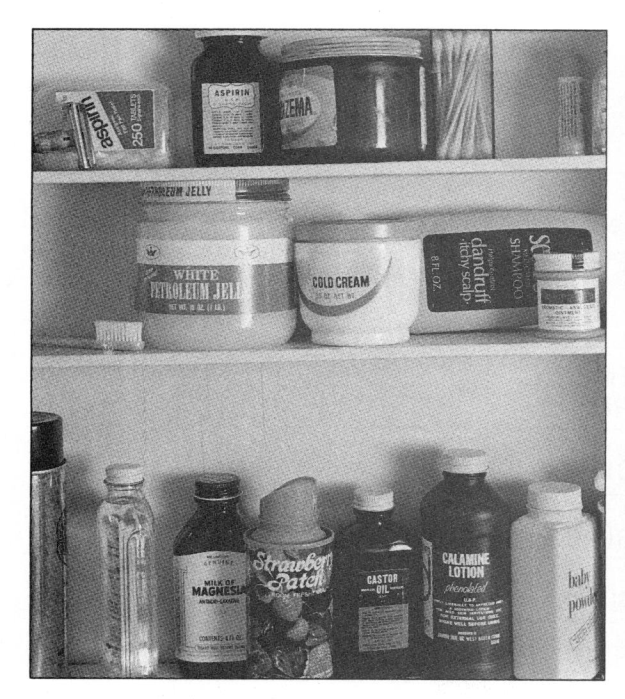

Figure E *Milk of magnesia is used to neutralize excess stomach acids.*

Figure F *Ammonium hydroxide is important in making fertilizers.*

FILL IN THE BLANK

Complete each statement using a term or terms from the list below. Write your answers in the spaces provided.

bitter	dangerous	opposite to
chemicals	do not change	pink
change	indicators	sour
lye		

1. Bases are a group of ________________ .
2. Bases often act________________________ the ways that acids act.
3. Both acids and bases can be ________________ .
4. Bases have a ________________ taste.
5. Acids have a ________________ taste.
6. Bases ________________ the color of red litmus paper.
7. Bases ________________________ the color of blue litmus paper.
8. Phenolphthalein turns________________ in bases.
9. Phenolphthalein and litmus paper are ________________ .
10. Soap contains a base called________________ .

MATCHING

Match each term in Column A with its description in Column B. Write the correct letter in the space provided.

	Column A	Column B
________	1. red litmus paper	a) ammonia
________	2. blue litmus paper	b) turns pink in bases
________	3. phenolphthalein	c) turns blue in bases
________	4. an acid	d) stays blue in bases
________	5. a base	e) vinegar

TRUE OR FALSE

In the space provided, write "true" if the sentence is true. Write "false" if the sentence is false.

________ **1.** Bases taste sour.

________ **2.** Bases feel slippery.

________ **3.** Bases turn blue litmus paper red.

________ **4.** Bases turn red litmus paper blue.

________ **5.** Phenolphthalein turns deep pink in bases.

________ **6.** Bases wear away metals.

________ **7.** Bases can be dangerous.

________ **8.** Acids contain the OH^- groups.

________ **9.** Acids contain the H^+ groups.

________ **10.** All bases are strong.

REACHING OUT

Why are indicators useful? ____

____

____

What happens when acids and bases are mixed?

18

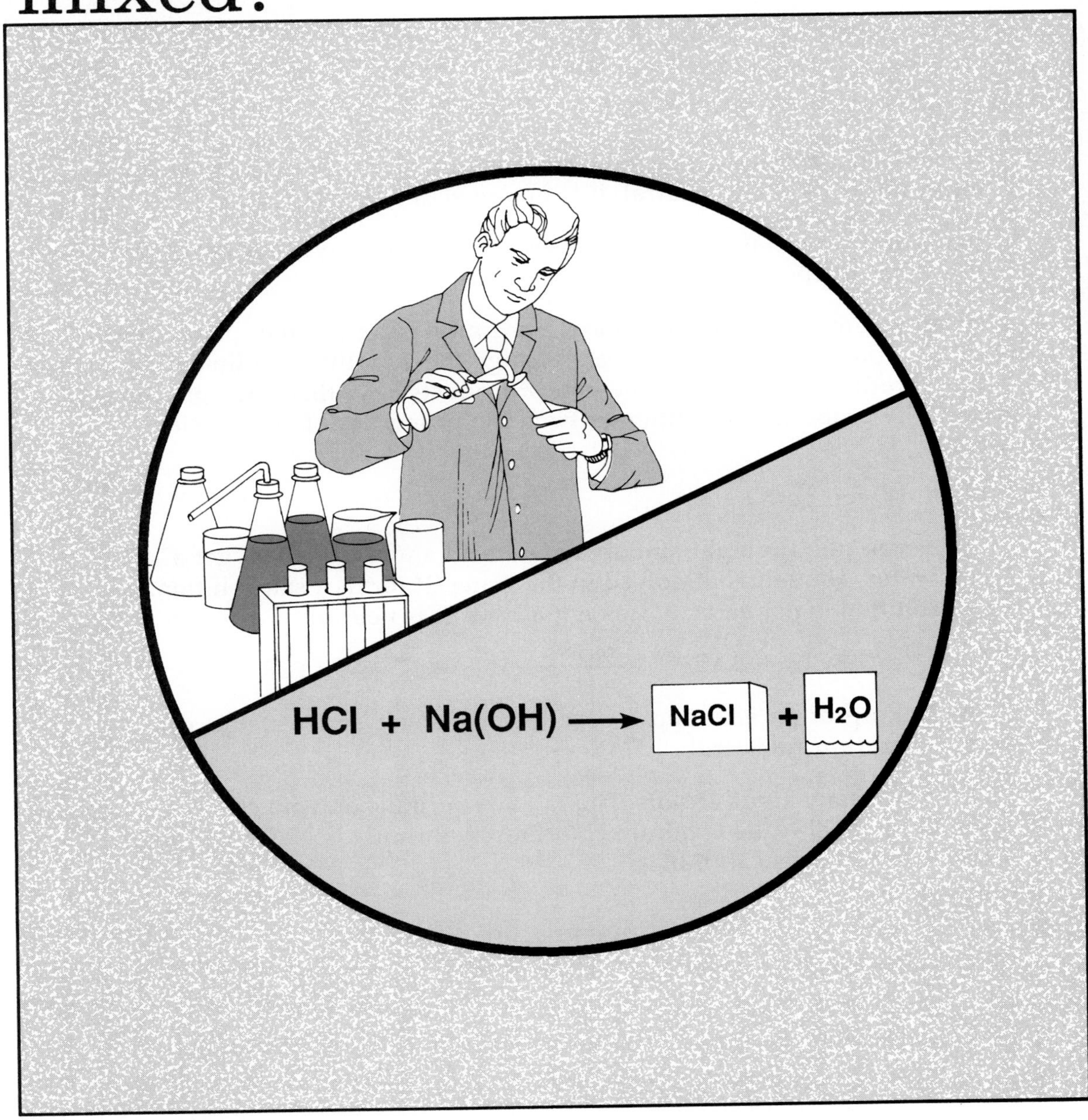

neutral: neither acidic nor basic

neutralization [new-truh-li-ZAY-shun]: reaction between an acid and a base to produce a salt and water

LESSON 18 What happens when acids and bases are mixed?

In chemistry, a liquid is **neutral** if it is not an acid nor a base. Take water, for example. Water is neutral. It is not an acid. It is not a base.

Acids and bases have definite properties. In many ways they are opposite What happens if you mix an acid with a base?

When you mix an acid with a base, a chemical reaction takes place. The atoms from the acid and the base change the way they are linked up. New products are formed. These new products have their own properties. The properties are different from the properties of either acids or bases.

What do you get?

When you mix the right amounts of an acid and a base, you get a salt and water. The salt is dissolved in the water. It forms a salt solution. A salt solution is not an acid: it is not a base. It is neutral.

$$\text{ACID + BASE} \xrightarrow{\text{makes}} \text{SALT + WATER}$$

The link-up of an acid and a base to form a salt and water is called **neutralization** [new-truh-li-ZAY-shun].

There are many kinds of salts. The salt you sprinkle on your food is just one kind of salt called sodium chloride. Its formula is NaCl. Different salts have different formulas.

MIXING AN ACID AND A BASE

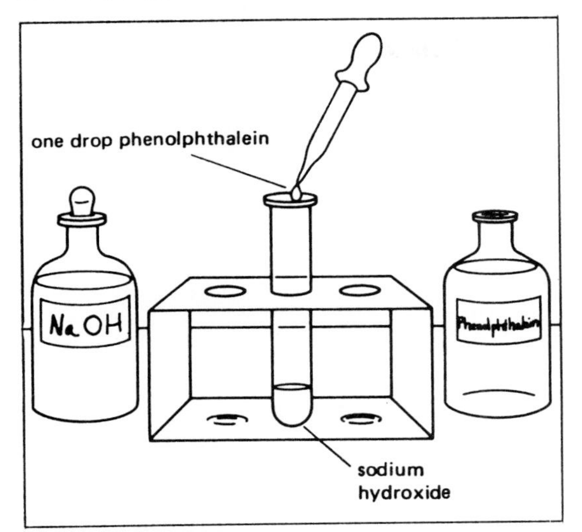

Figure A

The test tube in Figure A contains twenty drops of sodium hydroxide (NaOH).

One drop of phenolphthalein is added. The phenolphthalein turns deep pink.

This shows that sodium hydroxide (NaOH)

is ______________________ .
an acid, a base

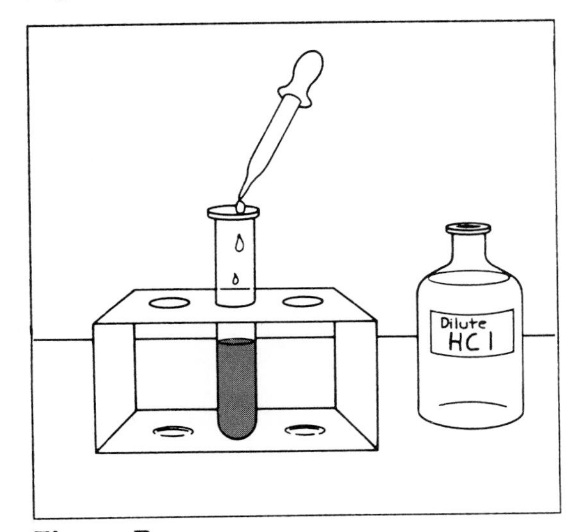

Figure B

A different dropper is used in Figure B to add fifteen drops of hydrochloric acid (HCl) — one drop at a time.

The solution stays pink.

This show that the solution

______________________ .
is neutral, is an acid, is still a base

Figure C

More hydrochloric acid (HCl) is added — one drop at a time, until the pink disappears.

The loss of the pink color shows that the

solution is ______________________ .
an acid, no longer a base

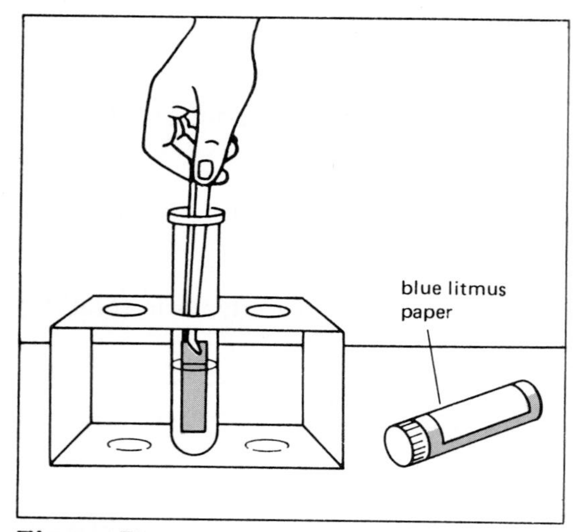

Figure D

The solution is tested with blue litmus paper.

The blue litmus paper stays blue.

This shows that the solution is not

____________________________ .
an acid, a base

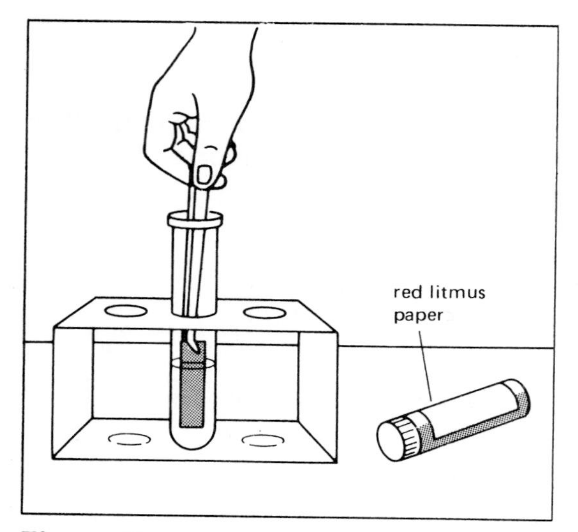

Figure E

The solution is tested with red litmus paper.

The red litmus paper stays red.

This shows that the solution is

____________________________ .
an acid, a base

The mixture ______________ neutral.
is, is not

Fill in the boxes to show what happened:

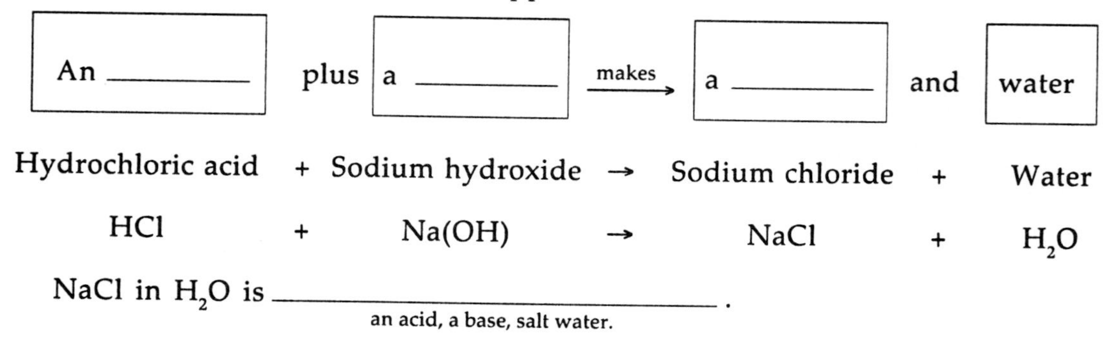

An ________ plus a ________ makes → a ________ and water

Hydrochloric acid + Sodium hydroxide → Sodium chloride + Water

HCl + Na(OH) → NaCl + H_2O

NaCl in H_2O is ________________________ .
an acid, a base, salt water.

FILL IN THE BLANK

Complete each statement using a term or terms from the list below. Write your answers in the spaces provided. Some words may be used more than once.

water	a base	table
neutralization	many kinds	neutral
litmus paper	reaction	phenolphthalein
an acid	a salt	

1. Lemon juice is an example of ______________ . Lye is an example of ______________ .
2. Any substance that is not an acid nor a base is said to be ______________ .
3. An example of a neutral liquid is ______________ .
4. The mixing of an acid and a base causes a chemical ______________ .
5. If we mix the right amounts of an acid and a base, we get ______________ and ______________ .
6. The chemical reaction between an acid and a base to produce a salt and water is called ______________ .
7. There are ______________ of salts.
8. The most common salt is ______________ salt.
9. Salt water does not change the color of ______________ or ______________ .
10. Salt water is neither ______________ nor ______________ . Salt water is ______________ .

MATCHING

Match each term in Column A with its description in Column B. Write the correct letter in the space provided.

	Column A	Column B
________	1. HCl	a) acid
________	2. NaOH	b) water
________	3. H_2O	c) base
________	4. NaCl	d) indicator
________	5. phenolphthalein	e) salt

TRUE OR FALSE

In the space provided, write "true" if the sentence is true. Write "false" if the sentence is false.

________ 1. An acid is neutral.

________ 2. A base is neutral.

________ 3. Water is neutral.

________ 4. There is only one formula for water.

________ 5. There is only one kind of salt.

________ 6. Salt water is neutral.

________ 7. If you mix an acid with a base, you get only water

________ 8. Blue litmus paper changes to red in salt water.

________ 9. Red litmus paper stays red in salt water.

________ 10. Phenolphthalein turns pink in salt water.

REACHING OUT

When hydrochloric acid reacts with potassium hydroxide, potassium chloride is formed.

- The formula for hydrochloric acid is HCl.
- The formula for potassium hydroxide is KOH.

What is the formula for water? ______________

What is the formula for potassium chloride? ______________

Why do some liquids conduct electricity?

19

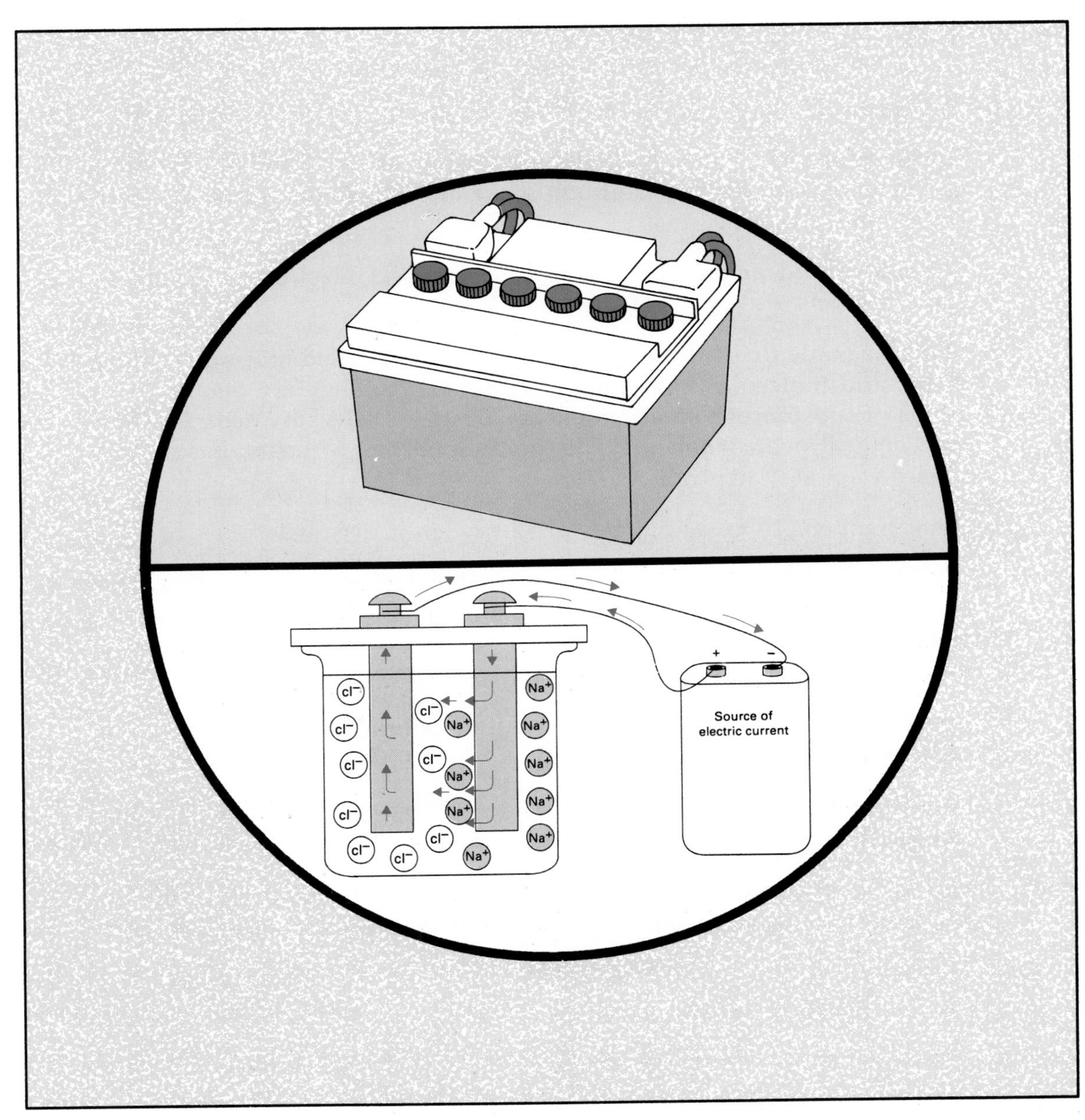

electrolyte [i-LEK-truh-lyt]: substance that conducts an electric current when it is dissolved in water

ion [Y-un]: charged particle

LESSON 19 | Why do some liquids conduct electricity?

Solutions of acids, bases, and salts conduct electricity. Solid acids, bases, and salts do not. Neither do liquids like alcohol, sugar water, distilled water, and glycerine.

Why do liquid acids, bases, and salts conduct electricity? Scientists explain it this way.

Matter is made up of atoms and groups of atoms called molecules. Most atoms and molecules have no electrical charge. They are neutral. The atoms of substances that are not acids, bases, or salts stay neutral. They stay neutral even when they dissolve. Solutions of acids, bases, and salts do not stay neutral.

What happens to an acid, base, or salt when it dissolves?

When an acid, base, or salt dissolves, its atoms do not stay together. The atoms unlock. When thy unlock, they do not stay neutral. They take on electrical charges. Some atoms have a positive (+) charge. Some have a negative (–) charge. Charged atoms are called **ions** [Y-unz].

Ions let electricity pass through a solution. Solutions that have ions are called **electrolytes** [i-LEK-truh-lyts].

Liquid acids, bases, and salts form ions. That is why they conduct electricity.

Liquids such as alcohol, sugar water, distilled water, and glycerine do not form ions. That is why they do not conduct electricity. They are nonelectrolytes.

TESTING ELECTROLYTES

We can test whether a solution conducts electricity by using a battery and a light bulb. The bulb lights up if the liquid is an electrolyte.

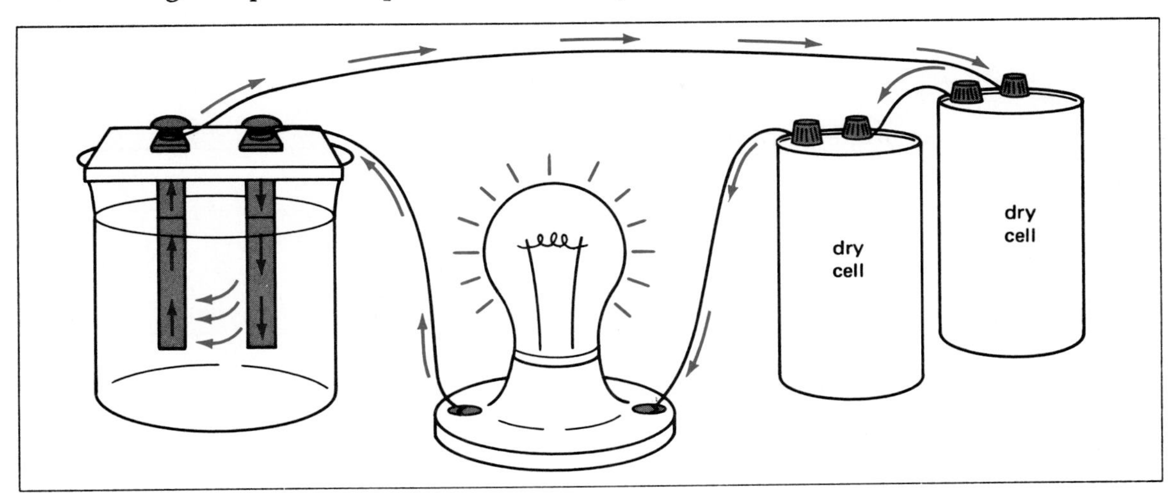

Figure A

The brightness of the light tells if the electrolyte is strong or weak.

- A bright light means a strong electrolyte (good conductor).
- A dim light means a weak electrolyte (poor conductor).
- No light means a nonelectrolyte (nonconductor).

The chart below lists 10 liquids. They have been tested as in Figure A. The checks show how brightly the bulb lit up.

	Liquid	Light Brightness		
		Bright	**Dim**	**No Light**
1.	Sodium chloride (salt)	✓		
2.	Sugar			✓
3.	Boric acid		✓	
4.	Sodium hydroxide (base)	✓		
5.	Distilled water			✓
6.	Acetic acid		✓	
7.	Alcohol			✓
8.	Magnesium sulfate (salt)	✓		
9.	Glycerine			✓
10.	Carbonic acid		✓	

Answer these questions about the information on the chart.

1. List the electrolytes ______________________________

____

____

2. a) Which are strong electrolytes? ______________________________

____

____

   b) How do you know? ______________________________

3. a) Which are weak electrolytes? ______________________________

____

   b) How do you know? ______________________________

4. a) Which are nonelectrolytes? ______________________________

____

   b) How do you know? ______________________________

5. Which groups of liquids let the bulb light up? ______________________________

____

6. Which groups of liquids are electrolytes? ______________________________

____

TRUE OR FALSE

In the space provided, write "true" if the sentence is true. Write "false" if the sentence is false.

________ 1. A regular atom has a charge.

________ 2. An ion has a charge.

________ 3. Only positive ions have a charge.

________ 4. Liquids that conduct electricity are called electrolytes.

________ 5. Electrolytes contain ions.

________ 6. An electrolyte contains plus and minus ions.

EXPLAINING IONIZATION

Let us see what happens in a salt solution.

Table salt has atoms of sodium (Na) and chlorine (Cl). One molecule of salt has one atom of sodium linked to one atom of chlorine. Atoms of sodium and chlorine have no charge. Molecules of the solid salt NaCl have no charge

NaCl in the solid state does not conduct electricity.

When NaCl dissolves, the sodium and chlorine atoms break away from each other. They take on electrical charges. The sodium takes on a positive charge. The chlorine takes on a negative charge.

The charged atoms are now called ions. Ions conduct electricity. Liquids that form ions are called electrolytes.

- Liquid acids, bases, and salts form ions.
- Liquid acids, bases, and salts are electrolytes.
- Liquid acids, bases, and salts conduct electricity.

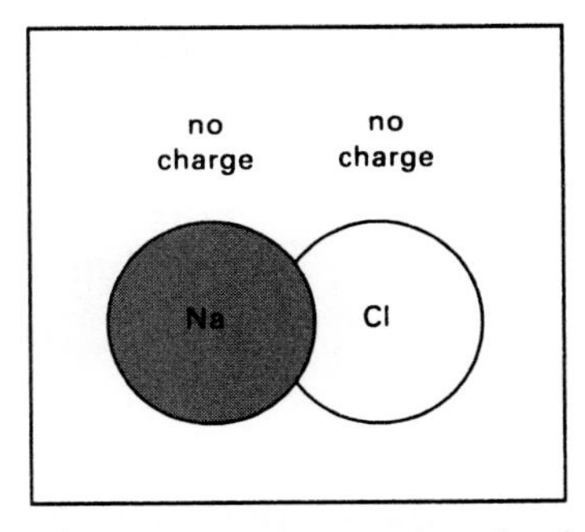

Figure B *One molecule of salt in the solid state*

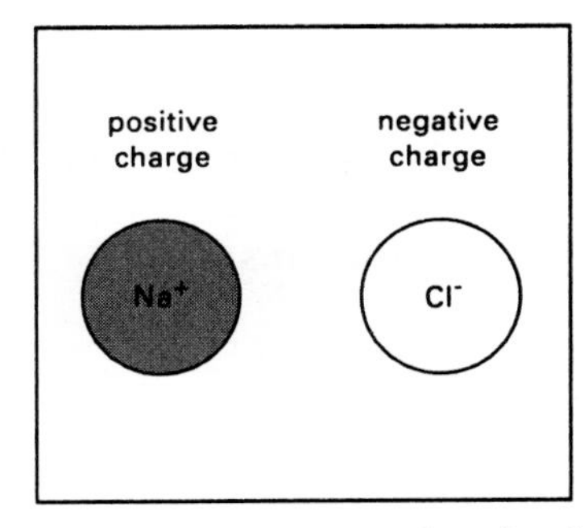

Figure C *One molecule of salt in the ionized state*

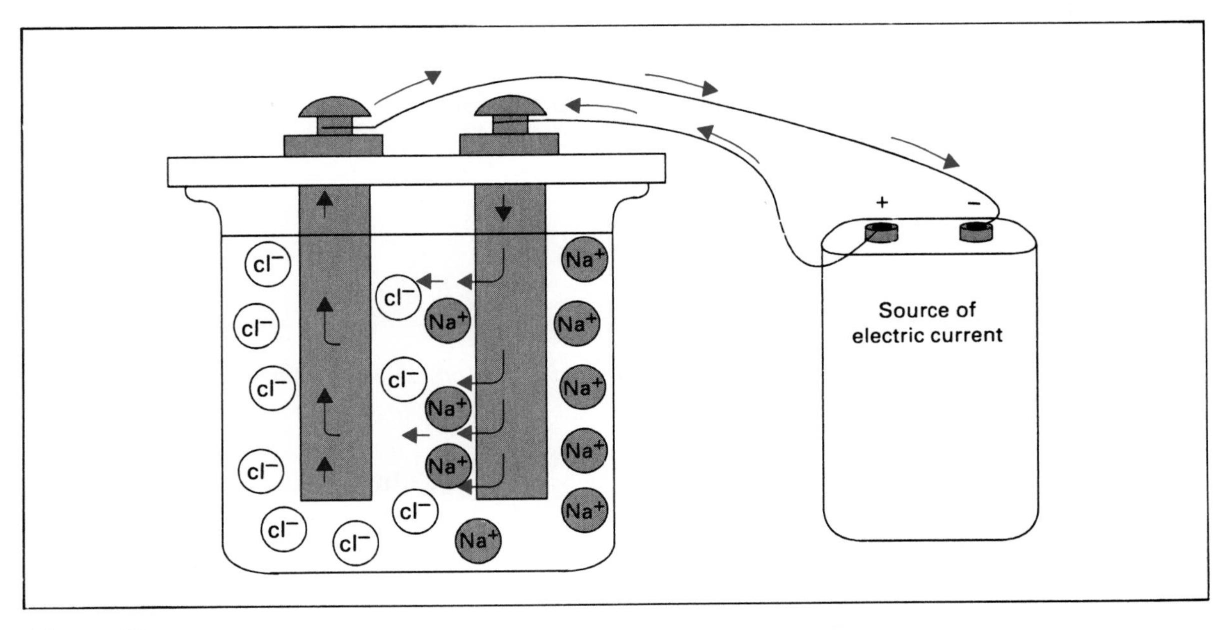

Figure D

FILL IN THE BLANK

Complete each statement using a term or terms from the list below. Write your answers in the spaces provided. Some words may be used more than once.

salts	electrolytes	negative
positive	no	atoms
ions	bases	conduct
acids		

1. Matter that is electrically neutral has ______________ charge.
2. Regular ______________ are electrically neutral.
3. Regular atoms have ______________ charge.
4. Atoms that have a charge are called ______________ .
5. Some ions have a ______________ charge; some have a ______________ charge.
6. Atoms of ______________ , ______________ , and ______________ form ions.
7. Ions ______________ electricity.
8. Solutions that conduct electricity are called ______________ .
9. Liquid______________ , ______________ , and ______________ are electrolytes.
10. Liquids like alcohol and distilled water do not form ______________ . They are not ______________ .

MATCHING

Match each term in Column A with its description in Column B. Write the correct letter in the space provided.

	Column A	Column B
________	1. ion	a) kinds of charges
________	2. positive and negative	b) do not conduct electricity
________	3. electrolyte	c) liquid that conducts electricity
________	4. regular atom	d) charged atom or molecule
________	5. dry acids, bases, and salts	e) has no charge

How do toxic wastes affect the environment?

20

pollutants [puh-LOOT-ents]: harmful substances
pollution [puh-LOO-shun]: anything that harms the environment

LESSON 20 | How do toxic wastes affect the environment?

Since the beginning of human history, people have been changing their environment. Some changes have been helpful. Others have been harmful. Floods, for example may cause a great deal of hardships, and even death. Dams have been built to control floods. This, of course, is a good change. However, dams also change the local environment in bad ways. Farmlands are left infertile. Plant and fish life in streams may be destroyed.

Pollution [puh-LOO-shun] is a harmful result of human activity. Pollution is anything that harms the environment. Pollution occurs when harmful substances called **pollutants** [puh-LOOT-ents] are released into the environment.

You probably know that pollution is a major problem. It harms every part of the environment, the air, the water, and the land.

Pollution is a worldwide problem. Cooperation among all nations is needed to help stop pollution.

Toxic wastes are one group of pollutants. Toxic wastes are poisonous chemicals and chemical by-products. Some are radioactive too. Radioactive substances are known to cause cancer and birth defects.

Factories, and especially chemical plants, produce most toxic wastes. What happens to the toxic wastes? Some are dumped into lakes, rivers and oceans. Others are buried in drums in the ground. In many places, the drums have rusted and broken apart. The toxic wastes are leaking from the drums into the ground. The wastes pollute the soil. They also seep into our water supply. Certain wastes have contaminated food, and water supplies. This has killed living things.

In 1980, a law was passed to clean up toxic waste sites in the United States. However, the clean-up of toxic wastes is a difficult problem. Cleaning up toxic wastes is expensive and takes a lot of time.

WHAT DO THE PICTURES SHOW?

Study Figures A, B, and C. Then answer the questions on the lines provided.

Figure A

Figure B

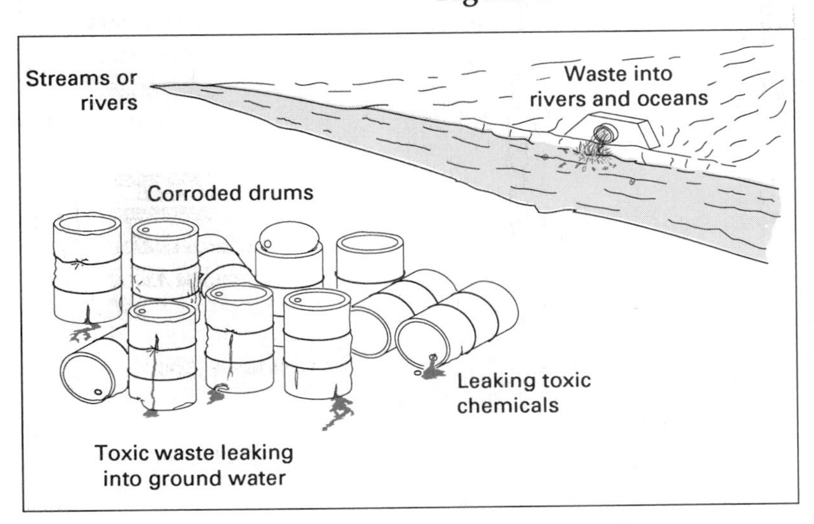

Figure C

1. Which figure shows toxic wastes being dumped directly into a stream? ________

2. a) What has happened to the drums in Figure C? ________________________

 ____

 b) What is happening to the toxic wastes in the drums? ________________

 ____

3. Why are the people in Figure B wearing protective suits? ________________

 ____

TOXIC WASTES

Trace the path toxic wastes may take in Figure D. Notice how they spread out in a web of damage and destruction. There are many possibilities.

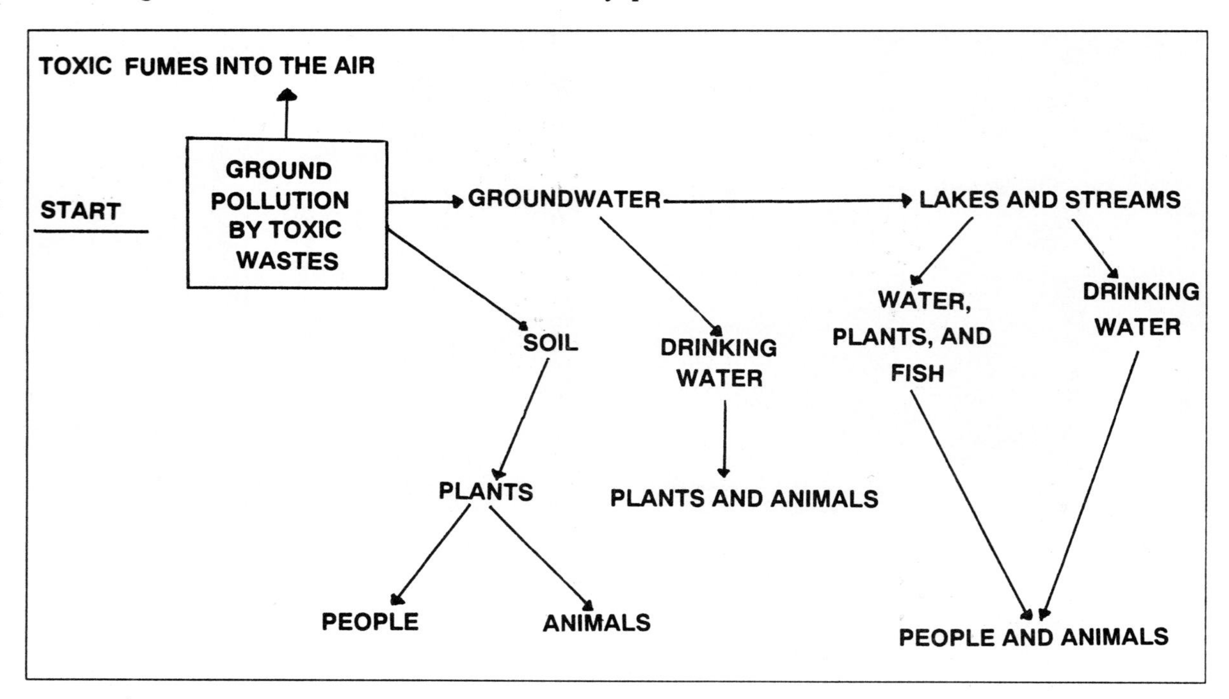

Figure D

Answer the questions about Figure D on the lines provided.

1. Toxic wastes can produce toxic ________________ that escape into the air.
2. Into what two things can toxic wastes leak? ________________ and ________________
3. Polluted ground water may be carried to ____________ and ________________.
4. What living things are directly affected by the pollution of lakes and streams? ________________________________
5. What living things are directly affected by soil pollution? ________________
6. Where does our drinking water come from? ________________________
7. What living things eat plants and fish that are affected by toxic chemicals? ________________________________
8. How do toxic wastes spread throughout the environment? ________________ ________________________________ ________________________________

TRUE OR FALSE

In the space provided, write "true" if the sentence is true. Write "false" if the sentence is false.

_________ **1.** Pollution is a worldwide problem.

_________ **2.** The clean-up of toxic wastes is easy.

_________ **3.** All toxic wastes are buried in the ground.

_________ **4.** Toxic wastes only harm fish and water plants.

_________ **5.** Pollution is anything that harms the environment.

_________ **6.** Radioactive substances cause cancer and birth defects.

_________ **7.** Toxic wastes do not produce fumes.

_________ **8.** Toxic wastes pollute water and food supplies.

_________ **9.** All human changes are bad.

_________ **10.** Chemical plants produce most toxic wastes.

NUCLEAR ENERGY

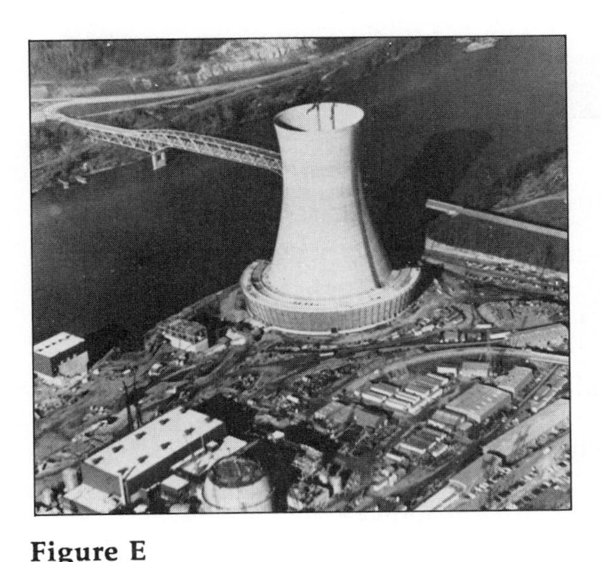

Figure E

Today people are using up oil, coal, and natural gas at a very rapid rate. These are our main energy resources. Therefore, people are looking for other energy sources.

Nuclear energy is another source of energy. Today, nuclear power plants have been built in many places. However, there are drawbacks to using nuclear energy. Dangerous radioactive wastes are produced. Storing and getting rid of these radioactive wastes is a major problem.

How can radioactive wastes harm people? ______________________________

____

____

CROSSWORD PUZZLE

Use the clues to complete the crossword puzzle.

Clues

Across

1. produce toxic chemicals
6. where toxic chemicals are buried
8. birth ______________
9. to fix
11. an energy resource
12. gases
14. poisonous chemicals

Down

2. chemically weathered
3. place
4. change
5. harmful substances
7. control floods
10. store toxic wastes
11. can be caused by radioactive substances
13. to leak

What is water pollution? 21

LESSON 21 | What is water pollution?

Smoke particles in the air, cloudy swirls of chemicals in a river—these are examples of suspensions. But these are special cases. The suspended particles are pollutants. Pollutants are harmful substances that are released into the environment.

Pollution is a major problem. Pollution is anything that harms the environment. Water pollution occurs when harmful substances enter the hydrosphere. Today, many lakes and rivers are polluted. They cannot be used for drinking or swimming. Some are so polluted that fish cannot live in them.

Now let us find out where water pollution comes from.

Sewage is a major source of water pollution. Germs live in sewage. Many fish and shellfish cannot be eaten because they contain germs that live in sewage.

Chemicals Many chemicals pollute the water. Some chemicals are used on farmlands to help plants grow. Others are used to kill insect pests. These chemicals seep into the ground water. The ground water is carried to lakes and rivers.

Some harmful chemicals come from industry too. Some factories dump wastes directly into rivers. Other industries bury wastes in drums in the ground. But what happens if the drums rust and break apart? The wastes leak into the ground water. Where do they end up?

WHAT DO THE PICTURES SHOW?

The pictures below show different causes of water pollution. Match each cause with its picture. Choose from the following causes:

Sewage
Oil spill
Industrial wastes
Farm chemicals

Figure A

1. Cause? ______________________

Figure B

2. Cause? ______________________

Figure C

3. Cause? ______________________

Figure D

4. Cause? ______________________

MORE ABOUT WATER POLLUTION

Figure E

Some industries take cold water from a lake or stream. They use the water for cooling. Then they release the water back into its river or lake. Is the water the same? No! When the water is used for cooling, the water itself becomes heated. So when the water is put back into its source, it kills plants and animals that normally live in cooler water.

This kind of pollution is called thermal [THUR-mul] pollution. Nuclear power plants are the main cause of thermal pollution.

TRUE OR FALSE

In the space provided, write "true" if the sentence is true. Write "false" if the sentence is false.

________ **1.** Pollution is not a major problem.

________ **2.** Some factories dump wastes directly into rivers.

________ **3.** Water pollution occurs when harmful substances enter the atmosphere.

________ **4.** Thermal pollution occurs when water is cooled.

________ **5.** Chemicals from farmland seep into ground water.

________ **6.** Germs live in sewage.

________ **7.** Pollutants of ground water are carried to lakes and rivers.

________ **8.** Farms are the major cause of thermal pollution.

________ **9.** Wastes buried in drums are not harmful to the environment.

________ **10.** Oil spills are harmful to ocean wildlife.

Figure F

Sewage-treatment plants have been built in many cities and towns. These plants change sewage into less harmful substances.

Figure G

Laws also help protect our water supply. Many farmland chemicals have been banned. They can no longer be used.

Laws require industries to clean their wastes before dumping them into lakes and rivers. Laws also call for the clean-up of wastes buried in drums in the ground.

1. Why is water pollution a major problem? ______________________

____

2. a) How do laws help fight water pollution? ______________________

____

b) What do you think should happen to people who break the law and illegally dump wastes into the water? ______________________

____

3. How can you help stop water pollution? ______________________

____

WORD SEARCH

The list on the left contains words that you have used in this Lesson. Find and circle each word where it appears in the box. The spellings may go in any direction: up, down, left, right, or diagonally.

- pollution
- water
- sewage
- germs
- thermal
- oil
- fish
- drums
- harmful
- wastes

P	A	J	R	L	G	S	H	P	S
W	O	K	L	E	J	H	M	L	E
A	I	L	R	P	I	S	N	U	W
S	L	M	L	O	C	I	A	F	A
T	S	G	B	U	O	F	Z	M	G
E	Y	F	X	R	T	B	E	R	E
S	N	I	E	F	D	I	R	A	D
M	U	T	Q	K	V	C	O	H	L
H	A	G	D	R	U	M	S	N	E
W	T	L	A	M	R	E	H	T	W

REACHING OUT

In the space below, draw a diagram that shows how a chemical used on a farm to kill insects could end up in a person's body.

How is water purified?

22

aeration [ayr-AY-shun]: spraying water into the air
chlorination [klor-uh-NAY-shun]: adding chlorine to the water
purified [PYOOR-ih-fyd]: cleaned

LESSON 22 | How is water purified?

The water you drink comes from wells, springs, or reservoirs [REZ-er-vwahrz]. Reservoirs are man-made lakes that store fresh water. This water is then piped into homes and businesses. What happens if pollutants fall into this water?

Before the water can be piped into homes and businesses, it must be cleaned, or **purified** [PYOOR-ih-fyd]. Water is purified in water-treatment plants.

There are five main steps in water purification. Think back to Lesson 3 where you learned how the parts of a suspension could be separated. Many of these methods are used to separate pollutants and other impurities from fresh water.

In one step, water is allowed to stand for a long time. In another step, chemicals are added to water. A third step includes filtering water.

Sometimes only one of these steps is used. However, usually more than one step is used. Look at Figure A on the next page as you read about each step.

HOW WATER IS PURIFIED

The water most people use in their homes has been treated in a water purification plant. Follow the different steps the water goes through to make it clean.

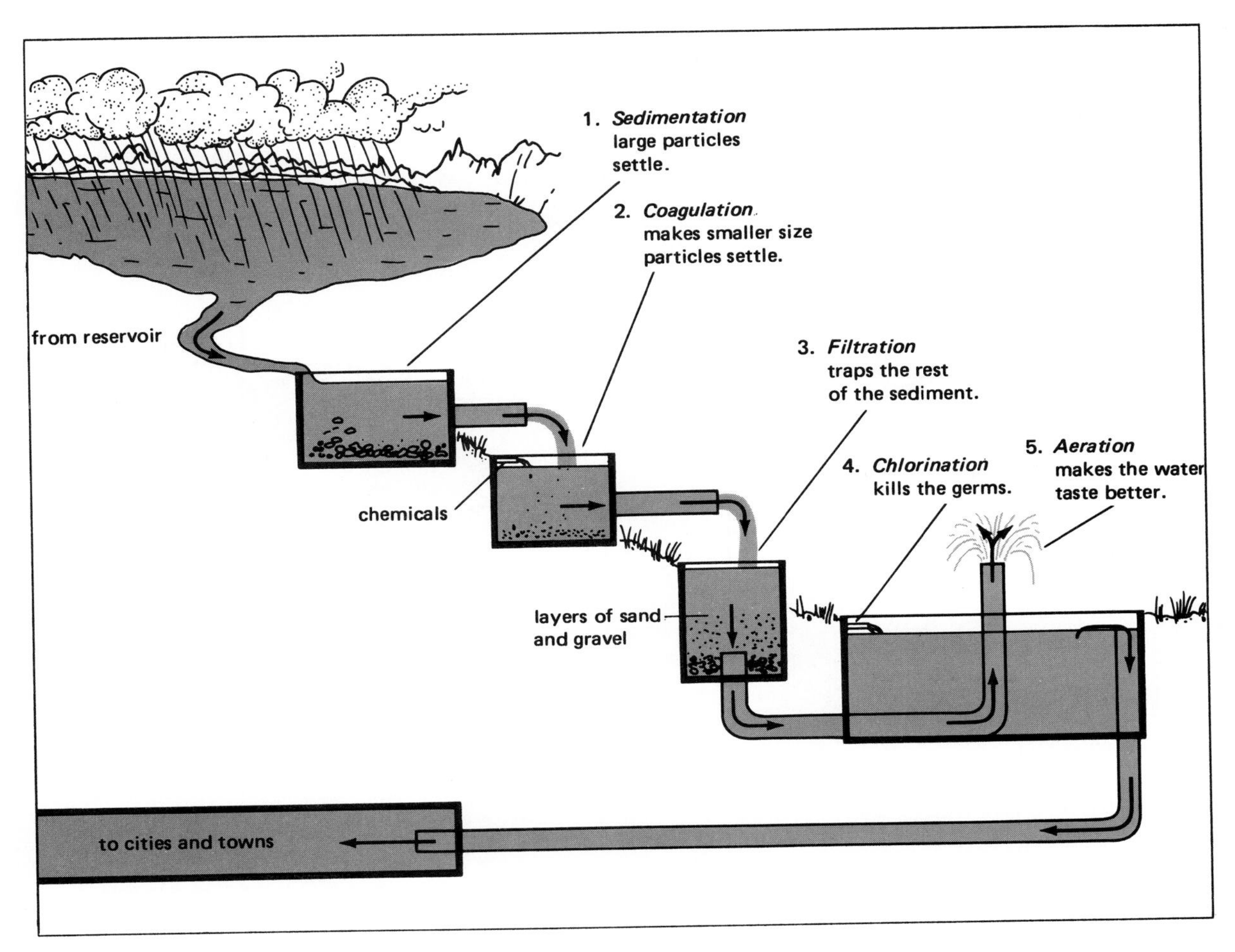

Figure A

Rain water is <u>usually</u> pure and safe. When rainwater hits the ground, it picks up sediment. Sometimes it even picks up germs and other impurities.

Step 1. Sedimentation [sed-uh-mun-TAY-shun]: water from the reservoirs is allowed to stand for long periods of time. Large and heavy particles settle to the bottom and are removed.

Step 2. Coagulation [koh-ag-yoo-LAY-shun]: small particles are still in the water after sedimentation. These smaller particles are made to settle by coagulation. Chemicals are added to the water. The chemicals make the particles lump together. They become heavy and settle.

Step 3. Filtration [fil-TRAY-shun]: water is passed through a filter to remove small particles. Filtration strains out any particles that are left after sedimentation and coagulation.

Step 4. Chlorination [klor-uh-NAY-shun]: chlorine is added to the water. The chlorine kills harmful microorganisms in the water.

Step 5. Aeration [ayr-AY-shun]: The water is sprayed into the air. The oxygen in the air dissolves in the water. The oxygen kills some additional microorganisms.

FILTER WATER YOURSELF

See for yourself how filtration takes out particles from water.

What You Need (Materials)

beaker of muddy water
funnel
ring stand
empty beaker
gravel
fine pebbles
sand

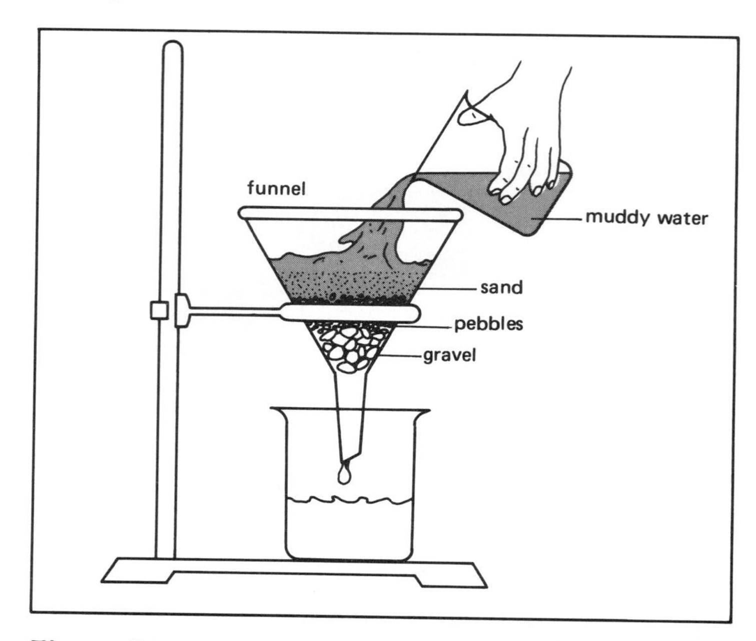

Figure B

How To Do the Experiment (Procedure)

1. Set up the funnel, ring stand, and empty beaker as shown in Figure B.
2. Pour some dirty water through the layers of sand and gravel.

What You Learned (Observations)

1. What does the water look like after it passes through the sand, pebbles, and gravel?

____

____

Something To Think About (Conclusion)

How could you make this water more pure? ______________________

____

____

FILL IN THE BLANK

Complete each statement using a term or terms from the list below. Write your answers in the spaces provided. Some words may be used more than once.

aeration	filtration
chlorination	sedimentation
coagulation	purification

1. Making water clean and safe to drink is called ______________ .
2. List the steps of purification in the order that they take place: ______________ , ______________ , ______________ , ______________ , ______________ .
3. Which step clears the heaviest particles? ______________
4. Which steps traps the last bits of dirt? ______________
5. Which step uses chemicals to help the pieces settle? ______________
6. Which step kills germs using chlorine? ______________
7. Which step uses oxygen in the air to kill microorganisms? ______________

TRUE OR FALSE

In the space provided, write "true" if the sentence is true. Write "false" if the sentence is false.

________ 1. Pollution has many sources.

________ 2. Water purification gets rid of sediment.

________ 3. Reservoirs hold water for drinking.

________ 4. Sediments can wash into reservoirs.

________ 5. Nothing has been done to stop pollution.

________ 6. Chlorination traps dirt.

________ 7. Aeration uses carbon dioxide to purify water.

________ 8. Coagulation uses chemicals.

________ 9. Gravity helps sedimentation.

________ 10. Only people in cities should worry about pollution.

SCIENCE *EXTRA*

Global Warming

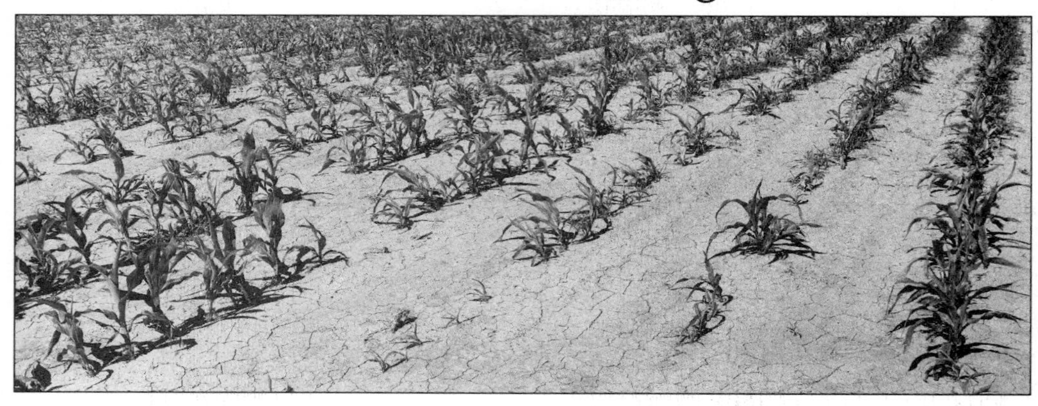

Each day, the sun warms the earth's surface. The heat, however, does not remain. The earth's surface emits heat back into the atmosphere, and then it slowly returns to space. An energy balance is created–ENERGY IN EQUALS ENERGY OUT. This keeps the atmosphere from becoming too hot, or too cold. Unfortunatedly, this delicate balance is being disturbed. In fact, it has been disturbed for a long time.

One of the gases present in the atmosphere is carbon dioxide (CO_2). Carbon dioxide forms a covering, similar to a blanket, over the earth. This blanket prevents heat loss. It prevents much of the heat from escaping the earth.

Modern society burns tremendous quantities of fossil fuels, oil, coal, and natural gas. The burning of these fossil fuels releases large amounts of carbon dioxide into the air. As a result, less heat escapes, and the air temperature rises.

Another contributing factor to the buildup of carbon dioxide in the atmosphere is the cutting down of trees. Trees and other plants use carbon dioxide to make food during photosynthesis. Plants and trees take in carbon dioxide and release oxygen. As more trees and forests get cut down, there are fewer trees that can use this carbon dioxide. Therefore, not as much carbon dioxide is being taken out of the air.

The continued rise in the level of carbon dioxide in the air and the expected rise in global temperature can have slow, but very serious effects. A higher temperature may result in the melting of glaciers and ice caps. This would raise the sea level, and flooding could occur. Climate, rain patterns and agriculture would change greatly. Some areas could become too hot or too dry to support agriculture.

How can the problem of global warming be slowed down? Global warming can be slowed by limiting the release of carbon dioxide and using fossil fuels more efficiently. Other solutions include using alternative energy sources, such as solar energy or wind energy. Another solution includes the planting of trees and the reforestation of forests. The problem of global warming will never be eliminated, but it can be reduced.

What is air pollution?

23

smog: mixture of smoke, fog, and chemicals

LESSON 23 | What is air pollution?

The burning of fossil fuels is the major cause of air pollution. Burning fossil fuels such as coal, oil, and natural gas releases poisonous gases into the atmosphere. Cars and factories use fossil fuels to run.

There are two kinds of air pollutants — solid particles and gases. Dust and soot are tiny particles. They are air pollutants. Dust and soot are given off in smoke. They can remain in the air for a long time. Dust and soot can irritate your eyes, and your lungs.

Many cities have a **smog** problem. Smog is a mixture of smoke, fog, and gas pollutants, such as carbon monoxide. Smog is harmful to people who have breathing problems.

Gas pollutants can also harm the environment indirectly. For example, sulfur dioxide is a pollutant gas. It combines with water in the atmosphere to form acids. The acids fall to the earth as acid rain. Acid rain damages statues and buildings. It also kills living things in lakes and streams. It kills trees too. You will learn more about acid rain in the next Lesson.

Many countries have laws to help control pollution. In the United States, cars must have anti-pollution devices. These devices prevent some pollutants that are given off by burning fuels, from entering the air. Factories also must have filters on their smokestacks.

AIR POLLUTION

Study the pictures below and read the text describing each picture. Then answer the questions.

Figure A

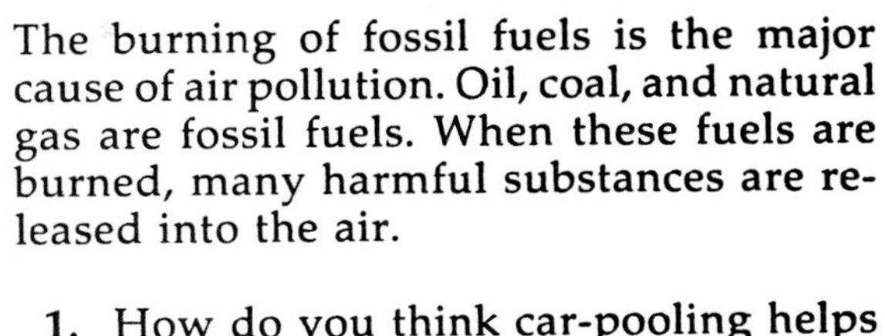

The burning of fossil fuels is the major cause of air pollution. Oil, coal, and natural gas are fossil fuels. When these fuels are burned, many harmful substances are released into the air.

1. How do you think car-pooling helps reduce air pollution? ______________________

____

____

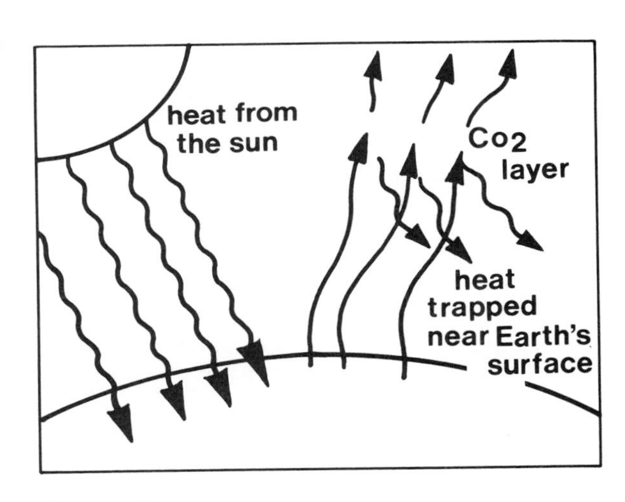

Figure B

Fuels need oxygen when they burn. They give off carbon dioxide. Carbon dioxide traps heat energy from the sun.

2. Scientists think that the increase of carbon dioxide in the air is causing the temperature of the earth to

____
rise, fall

3. What is the major cause of air pollution? ______________________

____

4. What are the three fossil fuels? ______________________

5. What are fossil fuels used for? ______________________

OBSERVING AIR POLLUTANTS

What You Need (Materials)

hand lens
microscope slide
toothpick
petroleum jelly

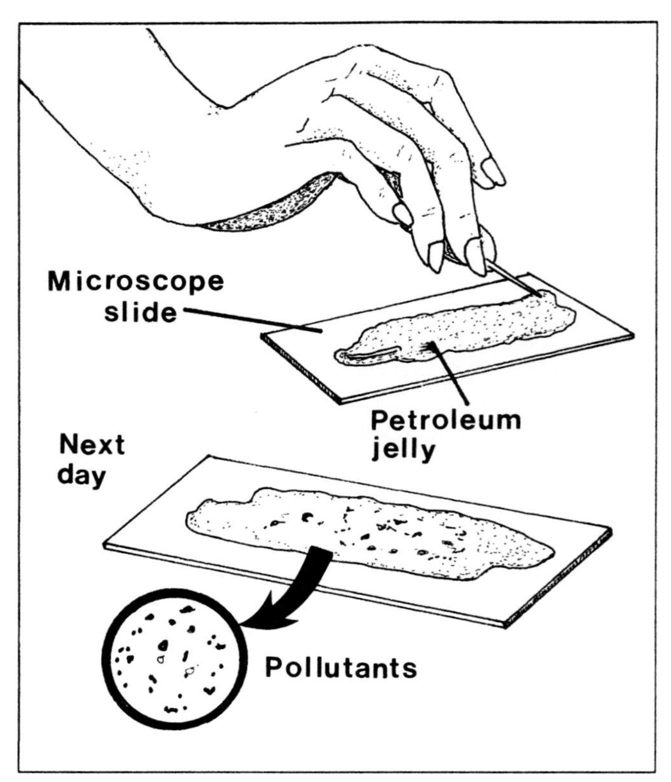

Figure C

How To Do the Experiment (Procedure)

1. Use the toothpick to coat one side of a glass slide with a thin layer of petroleum jelly.
2. Place the slide on a window ledge overnight.
3. Examine the slide with a hand lens the next day.

What You Learned (Observations)

1. What are some of the things you saw on your slide? ______________________________

____

Something To Think About (Conclusions)

1. What do you think these particles are? ______________________________
2. Where do you think they come from? ______________________________

FILL IN THE BLANK

Complete each statement using a term or terms from the list below. Write your answers in the spaces provided. Some words may be used more than once.

soot	acid rain	energy
fossil fuels	atmosphere	gas
cities	chemicals	rise

1. The major cause of air pollution is the burning of ______________ .
2. Air pollution occurs when harmful substances are released into the ______________ .
3. Two kinds of pollutants in smoke are dust and ______________ .
4. Smog is a mixture of smoke, fog, and ______________ .
5. Some gases that are released into the atmosphere combine with water in the air to form ______________ .
6. Carbon dioxide traps heat ______________ from the sun.
7. Many ______________ have a smog problem.
8. Carbon dioxide in the air is causing the temperature to ______________ .
9. Sulfur dioxide is a pollutant ______________ .
10. Cars use ______________ to run.

COMPLETE THE CHART

Complete the chart by identifying the pollutant or pollutants that are contained in each item described in the first column. Place a check mark in the correct column or columns.

	Source	Smoke	Dust	Soot	Chemicals
1.	Burning fossil fuels				
2.	Smog				
3.	Factory smokestacks				
4.	Automobile exhaust				
5.	Burning wood in a fireplace				

TRUE OR FALSE

In the space provided, write "true" if the sentence is true. Write "false" if the sentence is false.

________ **1.** Dust and soot are gas pollutants.

________ **2.** Smog is harmless.

________ **3.** Acid rain kills fish.

________ **4.** Smog is made up of smoke, fog, and chemicals.

________ **5.** Smog is a major problem in small towns.

________ **6.** The burning of fossil fuels is the major cause of air pollution.

________ **7.** Gas pollutants only harm the environment directly.

________ **8.** Carpooling does not reduce air pollution.

________ **9.** Dust and soot may remain in the air for a long time.

________ **10.** Air pollutants can irritate your eyes and lungs.

WORD SCRAMBLE

Below are several scrambled words you have used in this Lesson. Unscramble the words and write your answers in the spaces provided.

1. SGOM ____________________

2. TNPOULLTSA ____________________

3. IRA ____________________

4. BNGNRUI ____________________

5. OTOS ____________________

What is acid rain?

24

acid rain: rain containing nitric acid and sulfuric acid

LESSON 24 | What is acid rain?

In Lesson 23, you learned about air pollution. What happens to all of the pollutants that are found in the air? Some of these pollutants form **acid rain**. Acid rain can be in the form of any precipitation, including rain, snow, or sleet. Acid rain is one of the most serious environmental threats today. The problem not only affects large cities, but is slowly moving into rural areas as well.

What causes acid rain? Acid rain is caused by the release of sulfur and nitrogen gases into the air. Sulfur and nitrogen gas combine with water vapor, and fall to the earth as acid rain.

The gas that forms most acid rain is sulfur dioxide. Coal, gasoline, and oil contain sulfur. These fuels are burned for their energy. The burning of these fuels releases sulfur dioxide into the air. The gases mix with the water in the air and produce sulfuric acid. The following equation shows the production of acid rain:

$$SO_2 + H_2O \rightarrow H_2SO_3 \text{ (sulfurous acid)}$$

Sulfurous acid mixes with precipitation and forms acid rain.

Acid rain has polluted thousands of lakes and ponds. It has destroyed the wildlife that lives in these areas. Acid rain has also destroyed vast wilderness areas. Acid rain also affects the vegetation living in the area it falls upon.

Acid rain also affects buildings and industry as well. The acids in acid rain cause stone, brick and metal surfaces to wear away, or weather. Acid rain affects all parts of society in many parts of the world. For this reason, international cooperation is needed to solve this very serious problem.

MORE ABOUT ACID RAIN

Figure A *Acid rain destroys lakes, wildlife, and trees.*

Figure B *Acid rain also wears away stone. This statue has been damaged by acid rain.*

1. Which pollutant gases cause acid rain? ______________________________
2. How do most of these gases get into the atmosphere? ______________________________

____

3. Which gas is the greatest cause? ______________________________

Figure C *This is a coal burning power plant. Notice the pollution that is gushing from its smokestack.*

Figure D *Some sulfur and nitrogen pollution comes from natural sources. They include volcanoes, forest fires, and bacterial decay. But they are not as significant in the global acid rain problem.*

As you have already learned, chemicals can be divided into two groups — acids, and bases. We can say they are opposites.

Some acids, like soda water, are weak acids. Others, like vinegar and lemon juice, are stronger acids.

The strength of acids (and bases) are measured in number values called pH. The pH values range from 0 to 14.

Study the pH chart below. Then answer the questions below the chart.

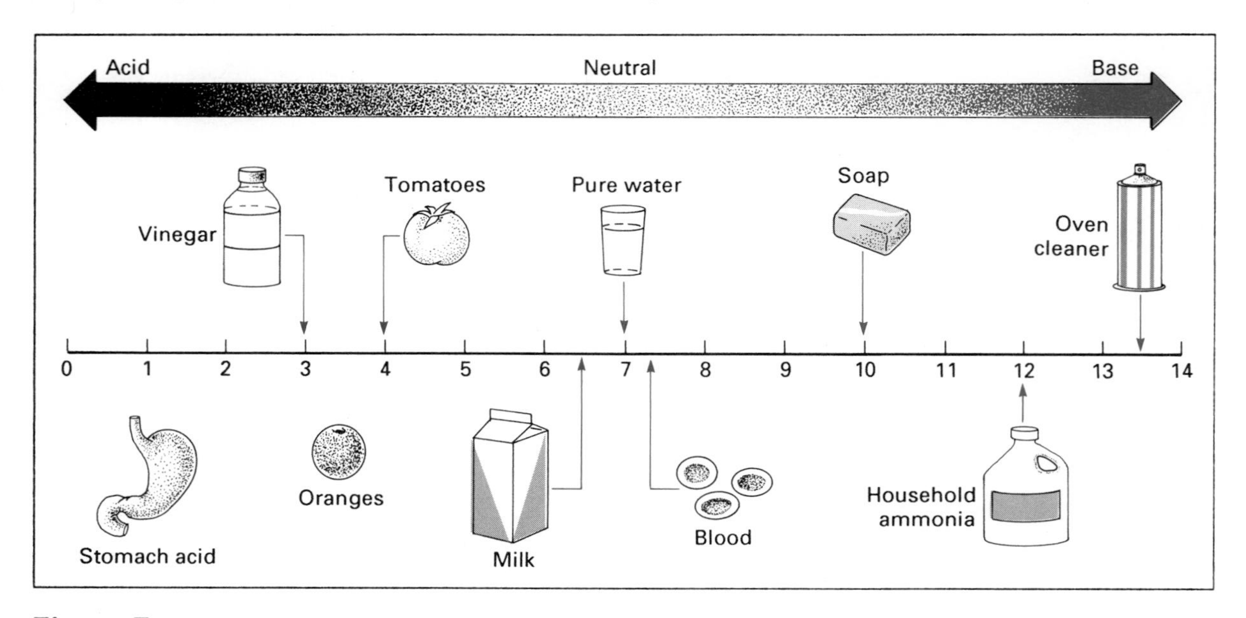

Figure E

1. Distilled water does not contain any chemicals. It is neither an acid nor a base. Distilled water is neutral.

   a) What is the pH of distilled water? ______________

   b) Chemicals with lower pH values are ______________ .
   acids, bases

   c) Chemicals with higher pH values are ______________ .
   acids, bases

2. a) The stronger an acid, the ______________ its pH value is.
   higher, lower

   b) The stronger a base, the ______________ its pH value is.
   higher, lower

ACID OR BASE?

Ten pH values are listed below. Identify each as an **acid,** *a* **base** *or* **neutral.** *Write the correct word next to each pH.*

1. 2.1 __________	**6.** 6.0 __________
2. 3.4 __________	**7.** 4.5 __________
3. 1.0 __________	**8.** 5.6 __________
4. 7.0 __________	**9.** 8.6 __________
5. 11.2 __________	**10.** 10.3 __________

Now look back to Figure E and answer the following questions.

What is the approximate pH value of . . .

1. vinegar ________

2. soap ________

3. milk ________

4. stomach acid ________

5. pure water ________

6. oranges ________

7. household ammonia ________

8. tomatoes ________

9. oven cleaner ________

10. blood ________

11. Which of the above items are acids? ____________________________

____

12. Which of the above items are bases? ____________________________

____

13. Which are neutral? ____________________________________

FILL IN THE BLANK

Complete each statement using a term or terms from the list below. Write your answers in the spaces provided. Some words may be used more than once.

acid	coal	oil	sulfur
acid rain	gasoline	pollution	snow
base	neutral	rain	water vapor
below	nitrogen	sleet	

1. Acid rain is caused by the release of ______________ and ______________ gases into the air.
2. Acid rain can be in the form of ______________ , ______________, ______________ or any other form of precipitation.
3. Most acid rain is caused by______________ gas.
4. Fuels burned for energy include ______________ , ______________ , and ______________ .
5. Any substance that harms the environment is______________ .
6. Sulfur and nitrogen gas combine with ______________ to produce ______________ .
7. A substance with a pH of 7 is said to be ______________ .
8. A substance with a pH of above 7 is said to be a ______________ .
9. A substance with a pH of below 7 is said to be an ______________.
10. Acid rain must have a pH of ______________ 7.

MATCHING

Match each term in Column A with its description in Column B. Write the correct letter in the space provided.

	Column A	Column B
________	1. acid	a) pollutant gas
________	2. base	b) substance with a pH of 7
________	3. neutral	c) substance with a pH below 7
________	4. acid rain	d) contains nitric acid and sulfuric acid
________	5. sulfur	e) substance with a pH above 7

WORD SEARCH

The list on the left contains words that you have used in this Lesson. Find and circle each word where it appears in the box. The spellings may go in any direction: up, down, left, right, or diagonally.

Acid
Acid rain
Air pollution
Base
Environment
Gases
Industry
Neutral
Nitrogen
Sulfur

N	O	E	N	J	U	N	E	T	H	E	T	W	E	L
I	T	I	N	D	U	S	T	R	Y	F	Y	T	A	N
T	R	S	T	V	A	L	B	E	R	T	A	R	N	S
R	D	I	W	I	I	L	L	C	L	O	T	E	S	U
O	E	A	C	I	D	R	A	I	N	U	A	S	N	L
G	E	C	W	H	O	M	O	E	E	W	I	A	L	F
E	L	I	B	E	D	U	R	N	S	I	R	B	E	U
N	A	D	L	L	Y	H	O	P	M	E	T	H	A	R
T	T	H	E	S	E	B	O	O	K	E	S	W	I	L
A	I	R	P	O	L	L	U	T	I	O	N	L	B	E
F	I	N	S	S	E	S	A	G	I	S	H	T	E	D

TRUE OR FALSE

In the space provided, write "true" if the sentence is true. Write "false" if the sentence is false.

________ **1.** Acid rain affects only living things.

________ **2.** Carbon dioxide is the main cause of acid rain

________ **3.** Acid rain causes bricks and metal to break down.

________ **4.** Sulfur is released by the burning of coal, oil, and gasoline.

________ **5.** Acid rain is a result of air pollution.

REACHING OUT

1. **Find out what is being done to reduce the effects of acid rain. Write a report on your findings.**

2. **In the space below, make a diagram to show the steps in the formation of acid rain.**

How does a nuclear reactor work?

25

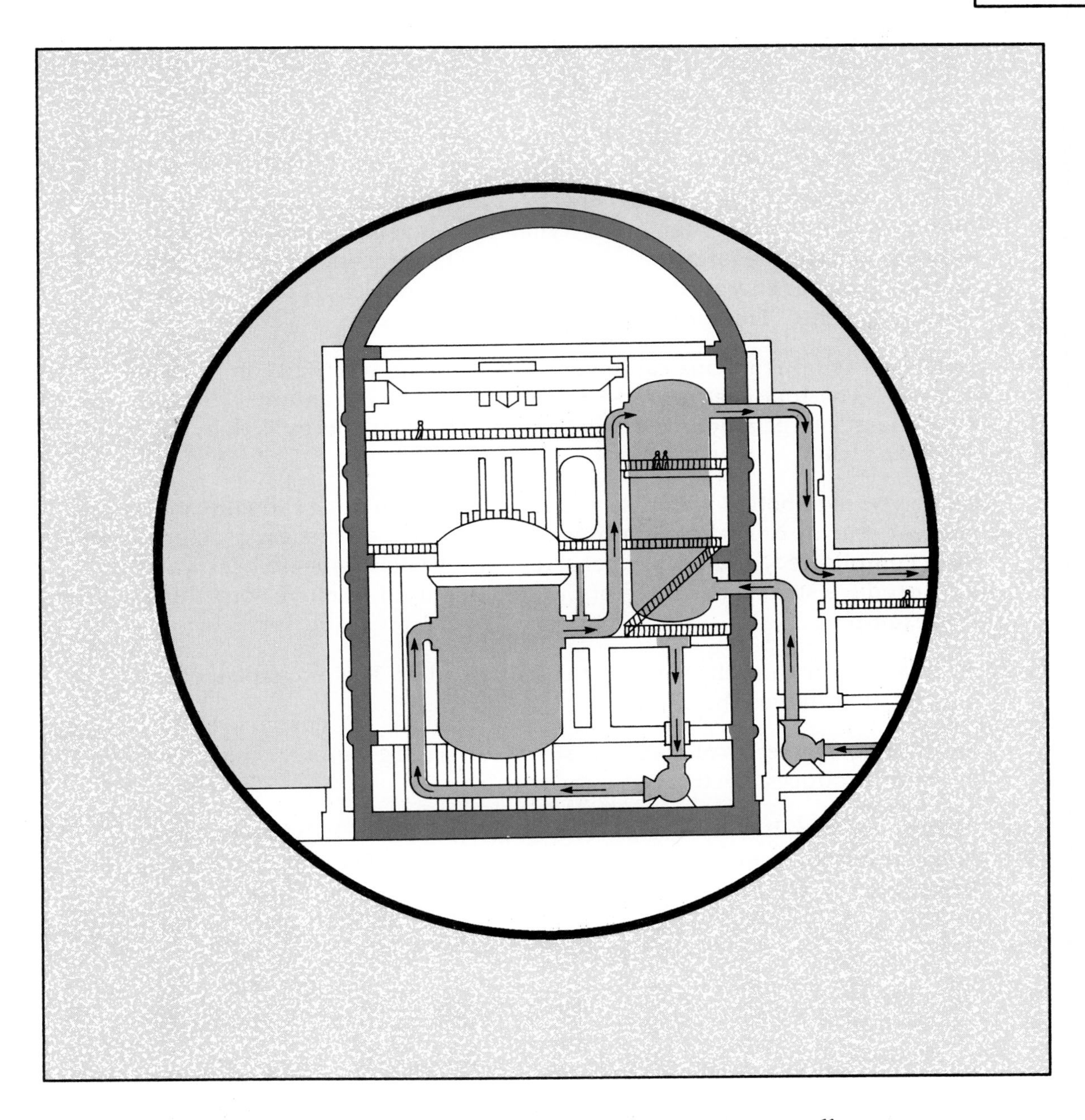

fission: nuclear reaction in which large atoms break apart into smaller atoms
fusion: nuclear reaction in which small atoms merge to form larger atoms
nuclear [NOO-klee-ur] **energy:** energy that is stored in the nucleus of an atom

LESSON 25 | How does a nuclear reactor work?

What is **nuclear** [NOO-klee-ur] **energy**? Nuclear energy is energy that is stored in the nucleus of an atom. When the nucleus is split, the energy is released as heat energy and light energy. This process is called nuclear **fission**. Nuclear energy also is related when the nuclei of atoms combine with each other. This process is called nuclear **fusion**.

Nuclear fission reactions can be used to generate electricity in nuclear reactors. The fuel used in most nuclear reactors is uranium. Uranium fuel releases large amounts of heat energy The heat energy is then used to make steam.

Have you ever boiled water? If you have, you have done the same job a nuclear reactor does, but in a much simpler way. A nuclear reactor is very complicated, and very expensive, but it has one job—to heat water and produce steam. The force of the steam is used to move something. For example:

- In a nuclear ship, the steam turns the propeller. The movement of the propeller moves the ship.
- In a nuclear electrical plant, the steam turns the generator. This movement produces electricity.

Nuclear energy is very useful. However it also can be very harmful. Nuclear energy poses special pollution problems. These problems will be covered in the next Lesson.

HOW A NUCLEAR REACTOR WORKS

Nuclear reactors are designed in many different ways. However, basically, they all have the same parts.

You are about to take a journey through a nuclear-reactor plant. Step-by-step, you will discover how the plant works.

Start by examining Figure A. Then you will study the plant — one part at a time. As it is taking form, you will learn the "what's and why's" of generating electricity with nuclear energy. Answer the questions as they appear.

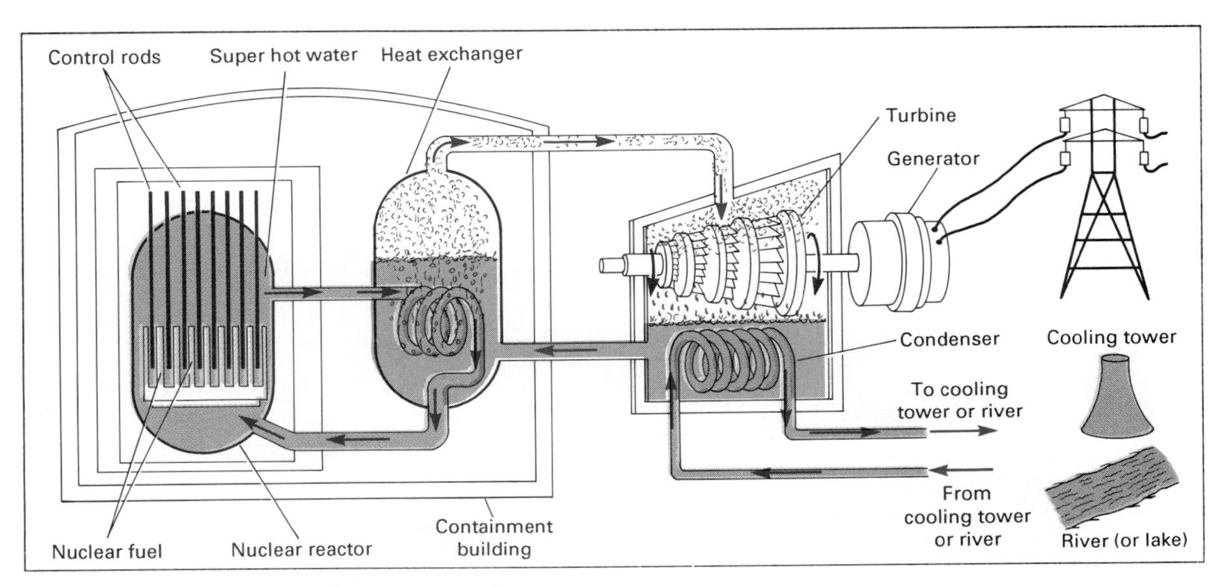

Figure A *Inside a nuclear reactor plant.*

PUTTING IT TOGETHER

The Reactor

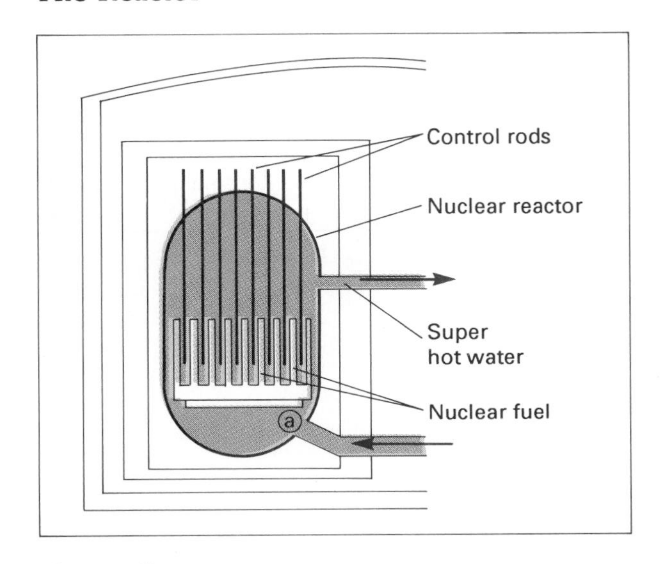

Figure B

1. The reactor contains cooled and superhot water. What else does the reactor contain? ______________________

____

2. The purpose of a nuclear reactor is to release heat. Where does this heat come from? ______________________

3. When control rods are lowered into the core, the rate of fission decreases. Control rods control the rate of the reaction. How do you think you could stop the reaction in an emergency? ______________________________

____

4. Cooled water enters the reactor through "a". (Actually, this water is not really "cool." It is warm.) What happens to this water in the reactor? ______________

____

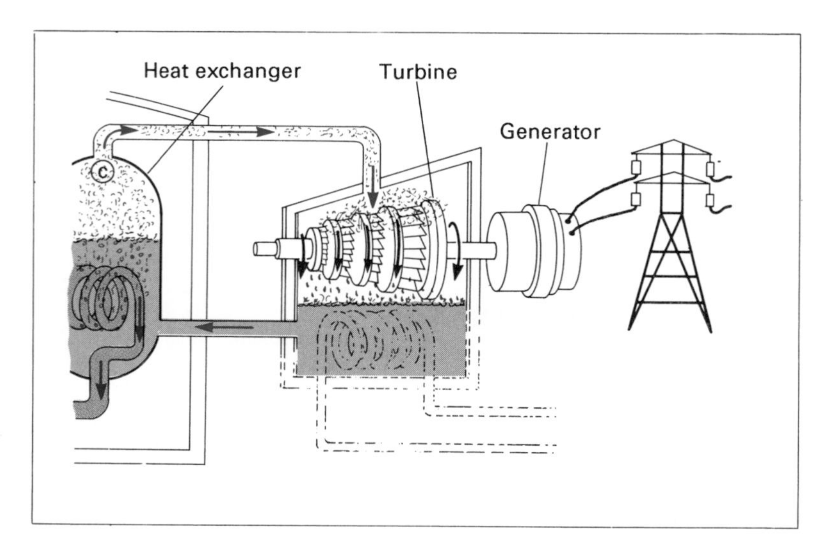

Figure C

5. From "b", the superhot water flows through pipes that enter into the ______ ______________ .

6. The heat exchange has its own water. What does the superhot core water do to the water of the heat exchange? It ______________ it, and changes it to ______________ .

Putting the Steam to Work

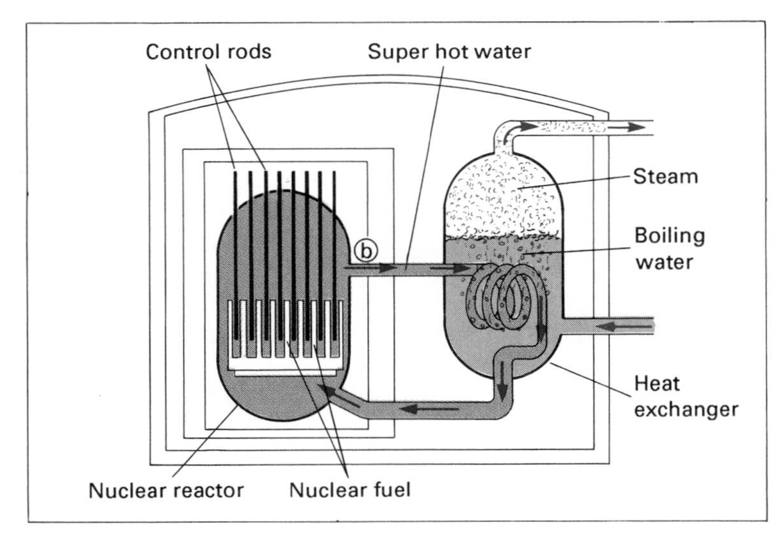

Figure D

Steam has power. It can make things move. THAT IS THE ENTIRE PURPOSE OF A NUCLEAR PLANT.

What happens next?

6. a) The steam leaves the heat exchanger through a pipe at _______________ .

   b) The pipe opens close to the blades of a _______________ .

   c) The escaping steam _______________ the turbine.

   d) The rotating turns the _______________ . This produces _______________ .

Is that it? No, there is one more step to go.

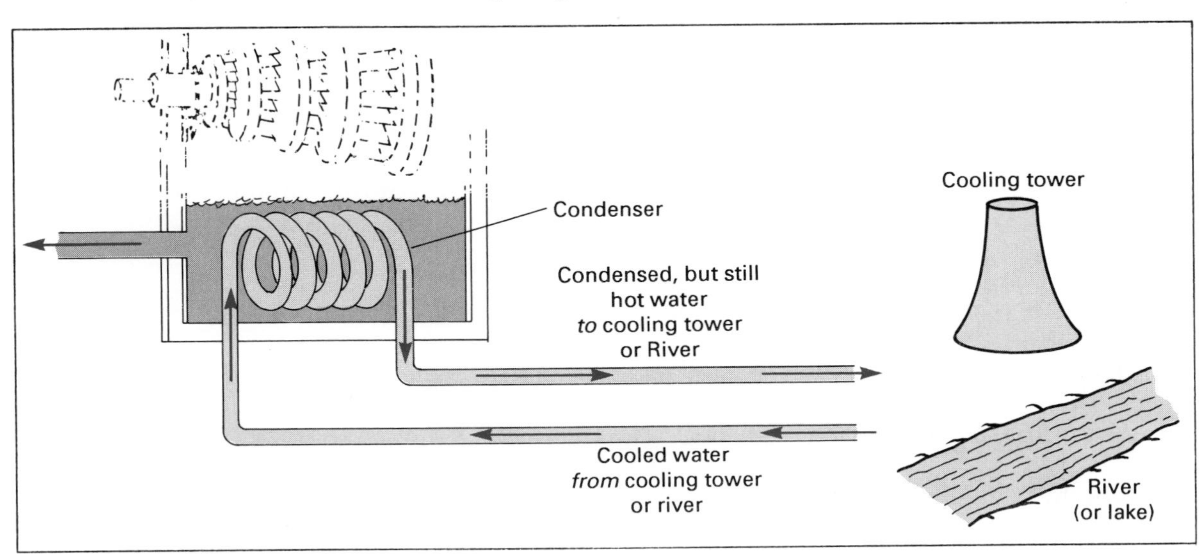

Figure E

The steam that turned the generator must be reused. But first, it must be condensed — changed back to a liquid.

7. a) This is done by pumping in _______________ water from a _______________

   or _______________ . This water flows in a separate pipe system.

   b) The cool water _______________ the environment around the steam.

   c) As a result, the steam condenses. What does this mean? _______________

   ____

FILL IN THE BLANK

Complete each statement using a term or terms from the list below. Write your answers in the spaces provided. Some words may be used more than once.

condensation	fusion	rotate
electricity	heat energy	uranium
fission	nuclear	steam

1. Energy that is stored in the nucleus of an atom is ______________ energy.
2. When two small atoms combine to form larger atoms, ______________ has taken place.
3. A nuclear reaction in which large atoms are broken apart into smaller atoms is called ______________ .
4. Nuclear fission is used to produce ______________ .
5. The fuel used in nuclear reactors is ______________ .
6. Uranium fuel releases ______________ .
7. The heat energy is used to produce ______________ .
8. Steam in a nuclear generator causes the turbine to ______________ .
9. The rotation of the turbine produces ______________ .
10. The steam is changed back to water by the process of ______________ .

WORD SCRAMBLE

Below are several scrambled words you have used in this Lesson. Unscramble the words and write your answers in the spaces provided.

1. FNSSIIO ______________________
2. ACTORRE ______________________
3. RANULCE ______________________
4. NIOFUS ______________________
5. DENNOCSE ______________________

What are nuclear wastes? 26

radioactive [ray-dee-oh-AK-tiv]: property of materials which gives off radiation from the nucleus of their atoms

LESSON 26 | What are nuclear wastes?

As you have just learned, nuclear energy is very important to society. There are advantages to nuclear energy as well as disadvantages. Some advantages of nuclear energy is that nuclear power plants use less fuel than other types of power plants. Another advantage is that uranium does not release chemical or other solid pollutants into the air.

The case against nuclear energy is just as strong. For example, nuclear power plants cost more money to build than other types of power plants. Another disadvantage is that nuclear power plants have the potential to be much more dangerous than other power plants. A third disadvantage is that uranium produces dangerous radiation even after it has been used up for energy. Nuclear wastes are mostly in the form of heat and radiation.

HEAT (THERMAL) POLLUTION Some waste heat vents directly into the air. Fortunately it poses no special problem. However, some waste heat is carried by water from the nuclear plant. It empties into a river or lake near the nuclear power plant.

Thermal water pollution raises the temperature of the water near the plant. This upsets the balance of the water. Life in and near the body of water is greatly affected. Some wildlife may die off.

RADIOACTIVITY All nuclear fuels are **radioactive** [ray-dee-oh-AK-tiv]. They give off very dangerous invisible rays. They can cause birth defects, burns, and severe illness.

Many safety measures are taken to guard against these dangers. The fuel itself is stored in special shielding containers. The nuclear reactor is housed in a thick concrete containment building.

The main problem lies in the "used" up nuclear fuel. Nuclear fuel that is no longer usable as fuel is still radioactive. The fuel may remain radioactive for hundreds of years.

In this lesson, you will learn of some of the ways of disposing of nuclear wastes.

THERMAL POLLUTION

Nuclear power plants usually are built near a water supply. You may recall from the last lesson that after the steam that turns the generator is used, it must be converted back into a liquid.

To condense the steam, it must be cooled off. The steam goes through cooling tanks or into a lake or river. Adding this heated water kills the wildlife living in the body of water. The heated water also may kill animals and plants living near the water supply.

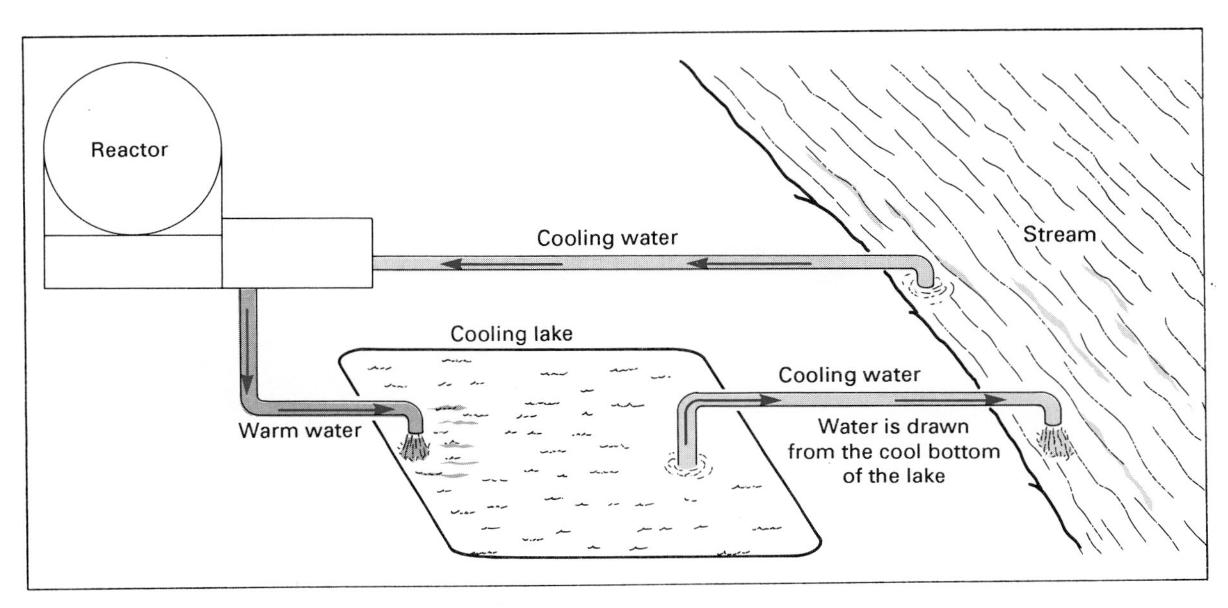

Figure A *Thermal pollution from a nuclear reactor.*

DISPOSAL OF NUCLEAR WASTES

Most nuclear wastes are placed in drums and buried deep in the ground. However, sooner or later, the drums will leak. What are some other suggestions for the disposing of radioactive waste?

- Radioactive wastes can be buried deep in the oceans.
- Radioactive wastes could be recycled and used again.
- Radioactive wastes could be stored in underground storage sites.

Figure B

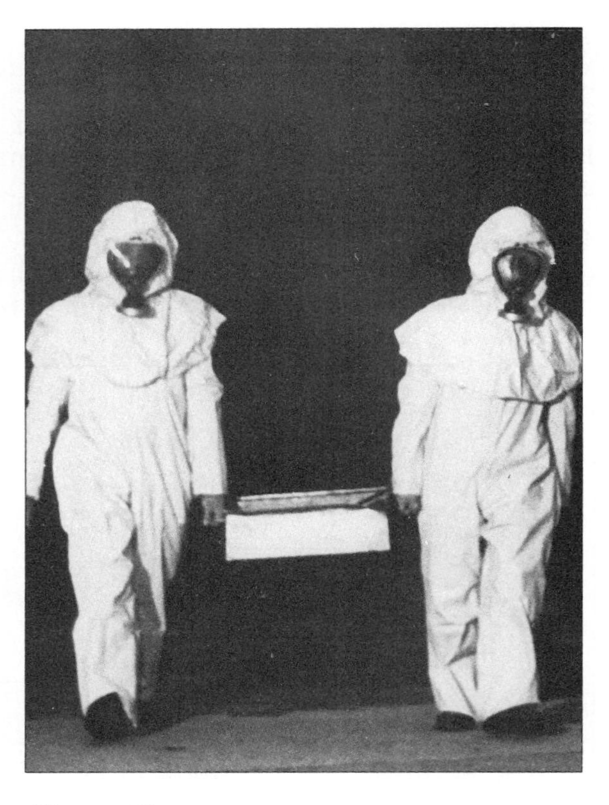

Figure C

Figure D

Figure E

MORE ABOUT RADIATION

Exposure to even a small amount of dangerous radiation can cause sickness and death. Radiation can come from natural as well as artificial sources. For example, ultraviolet rays from the sun is a natural source of radiation. Medical X-rays are examples of an artificial source of radiation.

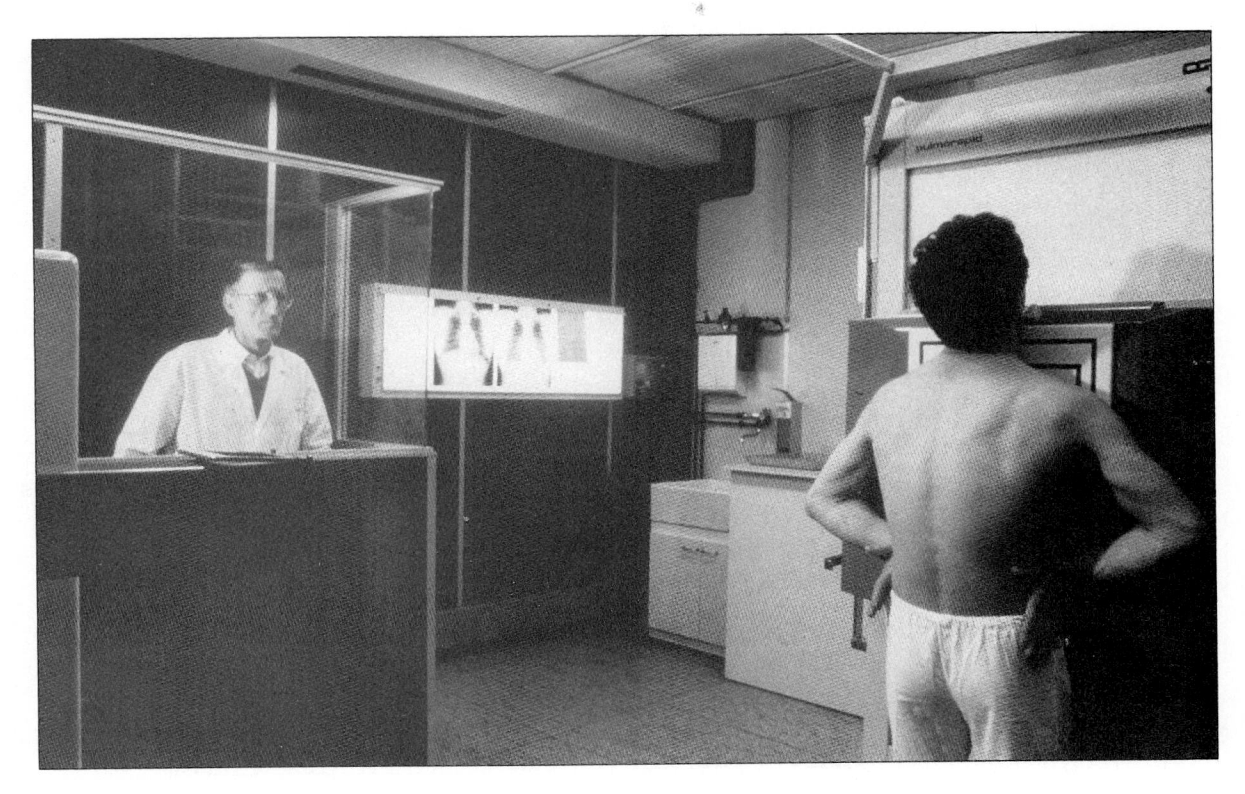

Figure F

Radiation is very useful in medicine. It is the energy source that makes X-rays possible. Properly controlled radiation can treat, and possibly cure, some illnesses such as cancer.

However, radiation can also cause cancer. Radiation damages the human immune system. Once the immune system, is damaged, it cannot recognize and fight disease. Radiation also damages cells. These cells rapidly divide and become cancerous.

The affects of radiation add up over time. Each time you are exposed to radiation, you increase the amount of radiation in your body. For this reason, you should limit your time in direct sunlight. Also, you should only get dental and medical X-rays if your doctor or dentist recommends them.

FILL IN THE BLANK

Complete each statement using a term or terms from the list below. Write your answers in the spaces provided. Some words may be used more than once.

heat	disposal	feel
cancer	radiation	thermal
temperature	nuclear reactor	uranium
hundreds		

1. Atomic energy is produced in a ______________________ .
2. Most atomic energy is released in the form of ________________ .
3. Most atomic pollution is in the form of extra ________________ and ________________ .
4. The term that refers to "heat" is ________________.
5. Nuclear thermal pollution upsets the _______________________ balance of bodies of water near nuclear plants.
6. The most common nuclear fuel is ________________ .
7. You cannot ________________ radiation.
8. Radiation can cause ________________ .
9. The great problem related to radioactive waste is ________________ .
10. Radioactivity — even of old, discarded atomic fuel, can last ________________ of years.

REACHING OUT

How do you think scientists could reduce the amount of thermal pollution produced at a nuclear power plant? ____

____

____

____

THE METRIC SYSTEM

METRIC-ENGLISH CONVERSIONS

	Metric to English	*English to Metric*
Length	1 kilometer = 0.621 mile (mi)	1 mi = 1.61 km
	1 meter = 3.28 feet (ft)	1 ft = 0.305 m
	1 centimeter = 0.394 inch (in)	1 in = 2.54 cm
Area	1 square meter = 10.763 square feet	1 ft^2 = 0.0929 m^2
	1 square centimeter = 0.155 square inch	1 in^2 = 6.452 cm^2
Volume	1 cubic meter = 35.315 cubic feet	1 ft^3 = 0.0283 m^3
	1 cubic centimeter = 0.0610 cubic inches	1 in^3 = 16.39 cm^3
	1 liter = .2642 gallon (gal)	1 gal = 3.79 L
	1 liter = 1.06 quart (qt)	1 qt = 0.94 L
Mass	1 kilogram = 2.205 pound (lb)	1 lb = 0.4536 kg
	1 gram = 0.0353 ounce (oz)	1 oz = 28.35 g
Temperature	Celsius = 5/9 (°F –32)	Fahrenheit = 9/5°C + 32
	0°C = 32°F (Freezing point of water)	72°F = 22°C (Room temperature)
	100°C = 212°F (Boiling point of water)	98.6°F = 37°C (Human body temperature)

METRIC UNITS

The basic unit is printed in capital letters.

Length	*Symbol*
Kilometer	km
METER	m
centimeter	cm
millimeter	mm

Area	*Symbol*
square kilometer	km^2
SQUARE METER	m^2
square millimeter	mm^2

Volume	*Symbol*
CUBIC METER	m^3
cubic millimeter	mm^3
liter	L
milliliter	mL

Mass	*Symbol*
KILOGRAM	kg
gram	g

Temperature	*Symbol*
degree Celsius	°C

SOME COMMON METRIC PREFIXES

Prefix		*Meaning*
micro-	=	0.000001, or 1/1,000,000
milli-	=	0.001, or 1/1000
centi-	=	0.01, or 1/100
deci-	=	0.1, or 1/10
deka-	=	10
hecto-	=	100
kilo-	=	1000
mega-	=	1,000,000

SOME METRIC RELATIONSHIPS

Unit	*Relationship*
kilometer	1 km = 1000 m
meter	1 m = 100 cm
centimeter	1 cm = 10 mm
millimeter	1 mm = 0.1 cm
liter	1 L = 1000 mL
milliliter	1 mL = 0.001 L
tonne	1 t = 1000 kg
kilogram	1 kg = 1000 g
gram	1 g = 1000 mg
centigram	1 cg = 10 mg
milligram	1 mg = 0.001 g

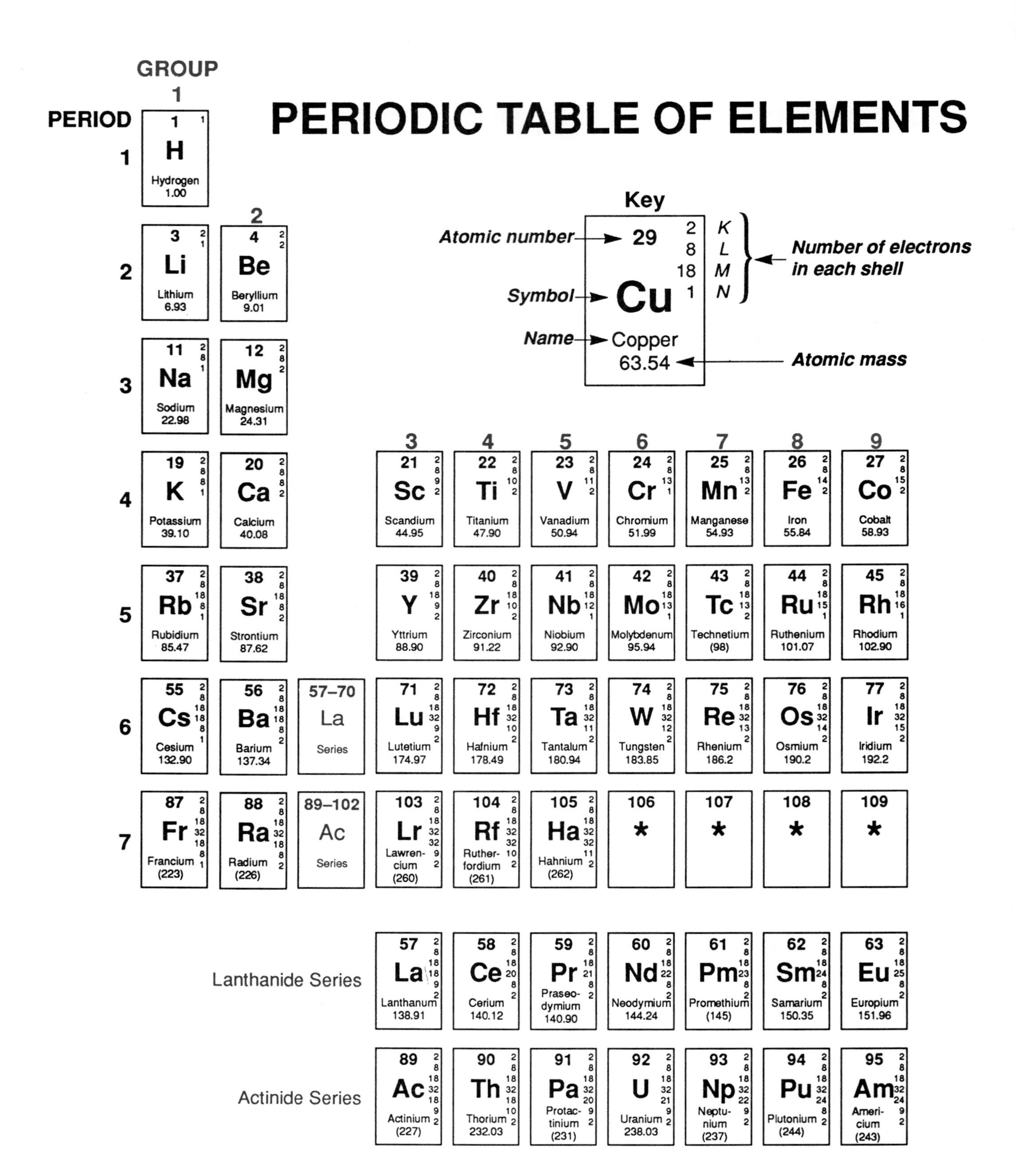

*** Names for these elements have not been agreed upon.**

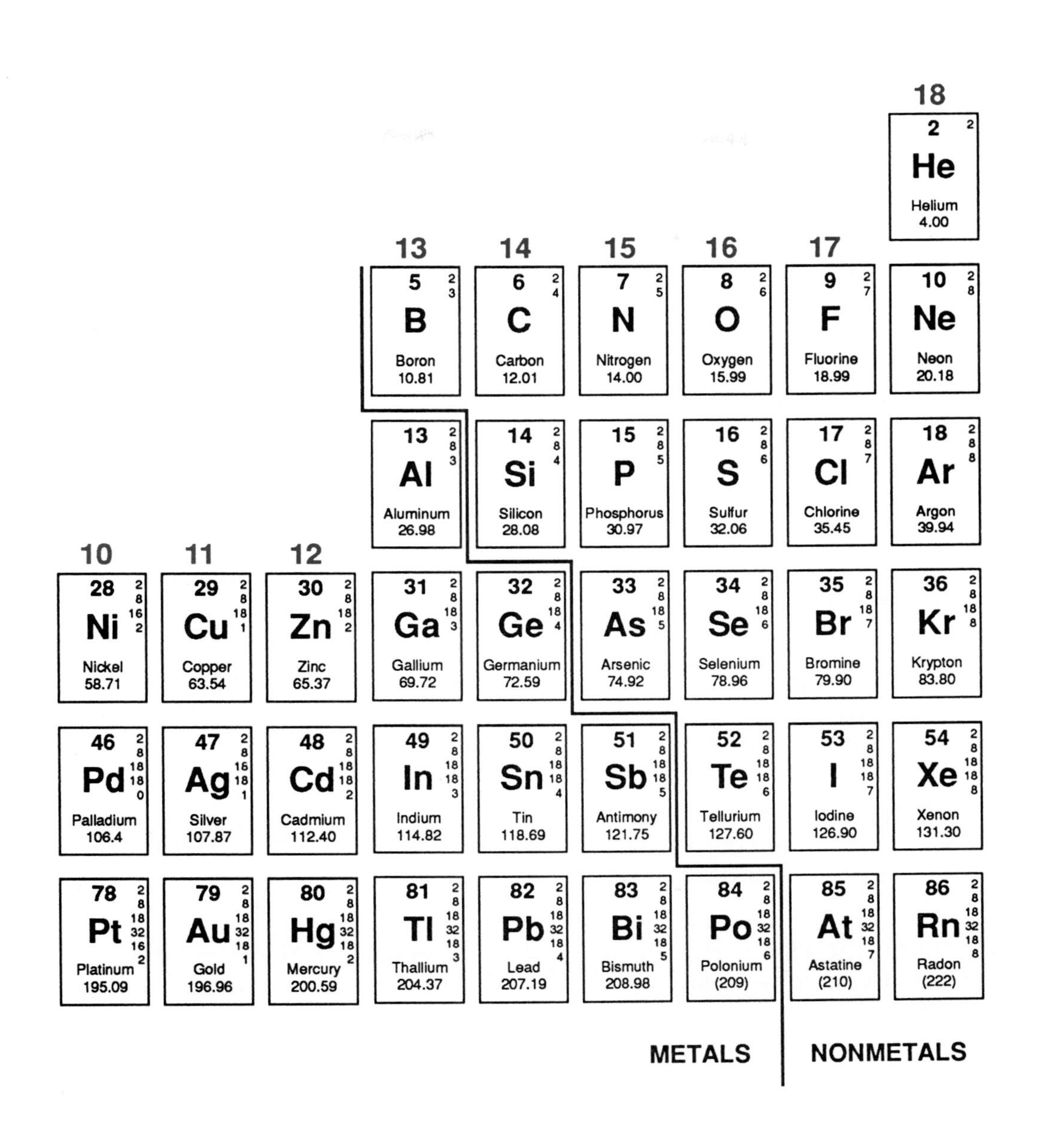
18
13
14
15
16
17
10
11
12
2 He Helium 4.00 2
5 B Boron 10.81 2 3
6 C Carbon 12.01 2 4
7 N Nitrogen 14.00 2 5
8 O Oxygen 15.99 2 6
9 F Fluorine 18.99 2 7
10 Ne Neon 20.18 2 8
13 Al Aluminum 26.98 2 8 3
14 Si Silicon 28.08 2 8 4
15 P Phosphorus 30.97 2 8 5
16 S Sulfur 32.06 2 8 6
17 Cl Chlorine 35.45 2 8 7
18 Ar Argon 39.94 2 8 8
28 Ni Nickel 58.71 2 8 16 2
29 Cu Copper 63.54 2 8 18 1
30 Zn Zinc 65.37 2 8 18 2
31 Ga Gallium 69.72 2 8 18 3
32 Ge Germanium 72.59 2 8 18 4
33 As Arsenic 74.92 2 8 18 5
34 Se Selenium 78.96 2 8 18 6
35 Br Bromine 79.90 2 8 18 7
36 Kr Krypton 83.80 2 8 18 8
46 Pd Palladium 106.4 2 8 18 18 0
47 Ag Silver 107.87 2 8 16 18 1
48 Cd Cadmium 112.40 2 8 18 18 2
49 In Indium 114.82 2 8 18 18 3
50 Sn Tin 118.69 2 8 18 18 4
51 Sb Antimony 121.75 2 8 18 18 5
52 Te Tellurium 127.60 2 8 18 18 6
53 I Iodine 126.90 2 8 18 18 7
54 Xe Xenon 131.30 2 8 18 18 8
78 Pt Platinum 195.09 2 8 18 32 16 2
79 Au Gold 196.96 2 8 18 32 18 1
80 Hg Mercury 200.59 2 8 18 32 18 2
81 Tl Thallium 204.37 2 8 18 32 18 3
82 Pb Lead 207.19 2 8 18 32 18 4
83 Bi Bismuth 208.98 2 8 18 32 18 5
84 Po Polonium (209) 2 8 18 32 18 6
85 At Astatine (210) 2 8 18 32 18 7
86 Rn Radon (222) 2 8 18 32 18 8
METALS
NONMETALS

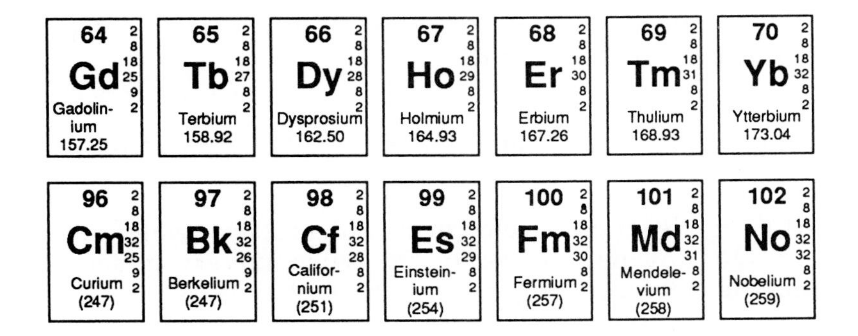
64 Gd Gadolin- ium 157.25 2 8 18 25 9 2
65 Tb Terbium 158.92 2 8 18 27 8 2
66 Dy Dysprosium 162.50 2 8 18 28 8 2
67 Ho Holmium 164.93 2 8 18 29 8 2
68 Er Erbium 167.26 2 8 18 30 8 2
69 Tm Thulium 168.93 2 8 18 31 8 2
70 Yb Ytterbium 173.04 2 8 18 32 8 2
96 Cm Curium (247) 2 8 18 32 25 9 2
97 Bk Berkelium (247) 2 8 18 32 26 9 2
98 Cf Califor- nium (251) 2 8 18 32 28 8 2
99 Es Einstein- ium (254) 2 8 18 32 29 8 2
100 Fm Fermium (257) 2 8 18 32 30 8 2
101 Md Mendele- vium (258) 2 8 18 32 31 8 2
102 No Nobelium (259) 2 8 18 32 32 8 2

SAFETY ALERT SYMBOLS

CLOTHING PROTECTION • A lab coat protects clothing from stains. • Always confine loose clothing.

EYE SAFETY • Always wear safety goggles. • If anything gets in your eyes, flush them with plenty of water. • Be sure you know how to use the emergency wash system in the laboratory.

FIRE SAFETY • Never get closer to an open flame than is necessary. • Never reach across an open flame. • Confine loose clothing. • Tie back loose hair. • Know the location of the fire-extinguisher and fire blanket. • Turn off gas valves when not in use. • Use proper procedures when lighting any burner.

POISON • Never touch, taste, or smell any unknown substance. Wait for your teacher's instruction.

CAUSTIC SUBSTANCES • Some chemicals can irritate and burn the skin. If a chemical spills on your skin, flush it with plenty of water. Notify your teacher without delay.

HEATING SAFETY • Handle hot objects with tongs or insulated gloves. • Put hot objects on a special lab surface or on a heat-resistant pad; never directly on a desk or table top.

SHARP OBJECTS • Handle sharp objects carefully. • Never point a sharp object at yourself—or anyone else. • Cut in the direction away from your body.

TOXIC VAPORS • Some vapors (gases) can injure the skin, eyes, and lungs. Never inhale vapors directly. • Use your hand to "wave" a small amount of vapor towards your nose.

GLASSWARE SAFETY • Never use broken or chipped glassware. • Never pick up broken glass with your bare hands.

CLEAN UP • Wash your hands thoroughly after any laboratory activity.

ELECTRICAL SAFETY • Never use an electrical appliance near water or on a wet surface. • Do not use wires if the wire covering seems worn. • Never handle electrical equipment with wet hands.

DISPOSAL • Discard all materials properly according to your teacher's directions.

GLOSSARY/INDEX

acid: substance that reacts with metals to release hydrogen, 98
acid rain: rain containing nitric acid and sulfuric acid, 146
aeration [ayr-AY-shun]: spraying water into the air, 136

base: substance formed when metals react with water, 104
boiling point: temperature at which a liquid changes to a gas, 74

chlorination [klor-uh-NAY-shun]: adding chlorine to the water, 136
coagulation [koh-ag-yoo-LAY-shun]: use of chemicals to make the particles in a suspension clump together, 14, 135
colloid [KAHL-oyd]: suspension in which the particles are permanently suspended, 22
concentrated [KAHN-sun-trayt-ed] **solution:** strong solution, 60
condensation [kahn-dun-SAY-shun]: change of a gas to a liquid, 92

dilute [di-LEWT] **solution:** weak solution, 60
dissolve: go into solution, 36
distillation [dis-tuh-LAY-shun]: process of evaporating a liquid and then condensing the gas back into a liquid, 92

electrolyte [i-LEK-truh-lyt]: substance that conducts an electric current when it is dissolved in water, 116
emulsifying [i-MUL-suh-fy-ing] **agent:** substances that keep an emulsion from separating, 28
emulsion [i-MUL-shun]: suspension of two liquids, 22
evaporate [i-VAP-uh-rayt]: change from a liquid to a gas, 74
evaporation [i-vap-uh-RAY-shun]: change of a liquid to a gas at the surface of the liquid, 92

filtration [fil-TRAY-shun]: separation of particles in a suspension by passing it through paper or other substances, 14, 135
fission: nuclear reaction in which large atoms break apart into smaller atoms, 154
freezing point: temperature at which a liquid changes to a solid, 80
fusion: nuclear reaction in which small atoms merge to form larger atoms, 154

homogeneous [hoh-muh-JEE-nee-us]: uniform; the same all the way through, 54
homogenization [huh-mahj-uh-ni-ZAY-shun]: formation of a permanent emulsion, 31

immiscible [i-MIS-uh-bul]: not mixable, 48
indicator [IN-duh-kayt-ur]: substance that changes color in acids and bases, 98
ion [Y-un]: charged particle, 116

miscible [MIS-uh-bul]: mixable, 45
mixture: two or more substances that have been combined, but not chemically changed, 2

neutral: neither acidic nor basic, 110
neutralization [new-truh-li-ZAY-shun]: reaction between an acid and a base to produce a salt and water, 110
nuclear [NOO-klee-ur] **energy:** energy that is stored in the nucleus of an atom, 154

phenolphthalein [fee-nohl-THAL-een]: an indicator that turns a deep pink color when a base is added, 104
pollutants [puh-LOOT-ents]: harmful substances, 122
pollution [puh-LOO-shun]: anything that harms the environment, 122
properties [PROP-ur-tees]: characteristics used to describe a substance, 54
purified [PYOOR-ih-fyd]: cleaned, 134

radioactive [ray-dee-oh-AK-tiv]: property of materials which gives off radiation from the nucleus of their atoms, 160

saturated [SACH-uh-rayt-id] **solution:** solution containing all the solute it can hold at a given temperature, 60

sedimentation: method of separating a suspension in which gravity makes solid pieces settle to the bottom, 14, 135

smog: mixture of smoke, fog, and chemicals, 140

solute [SAHL-yoot]: substance that is dissolved in a solvent, 36

solution: mixture in which one substance is evenly mixed with another substance, 36

solvent: substance in which a solute dissolves, 36

spinning: method of separating a suspension in which the outward force of spinning the container forces solid pieces to the bottom of the container, 14

suspension [suh-SPEN-shun]: cloudy mixture of two or more substances that settle on standing, 8

tincture [TINK-chur]: solution of a substance in alcohol, 50

transparent [trans-PER-unt]: material that transmits light easily, 54

water vapor: water in the gas state, 74